Hive and Seek

Rebecca O'Bea

Cover art by Socorro Reyes

Dedication

This book is dedicated to the memory of my father, LTC(ret.) Robert D. Tilley
A member of the greatest generation and the person who encouraged my love of reading.
★ A huge shoutout to my family and friends, who inspire and motivate me every day. Without them, this book would not have been possible. A special thanks to my husband, Michael, who has put up with my "crazy ideas" for more than thirty years. I love you!

Epigraph

Even as a swarm of bees, that sinks in flowers
One moment, and the next returns again
To where its labour is to sweetness turned,
Sank into the great flower, that is adorned
With leaves so many, and thence reascended
To where its love abideth evermore

Dante's Paradiso: Canto XXXI

Contents

Chapter One

- -

T he gears grind as I pop the clutch of the '69 Chevrolet step-side in my second attempt to shift down from third. Finally, the truck lurches forward, as does Karl Bauman, my beekeeping mentor, who sits opposite the bench seat. He braces one hand against the dashboard, and his back is straight. He is unhappy about my clumsy attempt to drive his beloved truck.

"You told me you knew how to drive a standard transmission," Karl says, a pained expression on his face.

"I do. I drove a stick shift in high school, a five-speed. Not this––" I wave at the three-on-the-tree gearshift from hell. Karl may call his truck a classic. But I'm convinced if a vehicle has a personality, this one's been possessed by a demon.

"Gertie takes a gentle hand," Karl says. "I told you to ease off the clutch."

"Gertie! Gertie drives like a tank. Besides, who names their fifty-year-old truck 'Gertie'?"

"My mother-in-law's name was Gertrude. She was a God-fearing woman, and everyone called her Gertie. Besides, you have no room to talk. What kind of name is 'Indigo Bleu'?"

He had me there. It was my given name for reasons my mother has yet to explain, and I still couldn't figure it out. "Well? What did your mother-in-law think of you naming a truck after her?"

1

"Oh, she hated it till the day she died. God rest her soul," Karl says, a sly grin across his face.

We merge at the four-way intersection of K5 and Mueller Road with no further incidents of user error. Let's face it: A four-way stop sign is nothing more than an exercise in indecision—an object lesson in critical thinking. One designed to bring out a driver's worst tendencies.

Of course, Gertie stalls on the slope and starts a slow roll backward. In the rearview mirror, I waggle my fingers at the slack-faced driver of the county co-op truck. He does not reciprocate.

To the west, a middle-aged blonde in a spandex athletic top gestures impatiently. A rolled-up yoga mat beside her and a pale pink shower scrub dangles from the driver's mirror. I recognize Brenda Matthews, the wife of the mayor of our tiny hamlet of Colette, Kansas, and she is obviously in a hurry to get somewhere. Brenda pounds on the horn of her silver SUV and mouths a word I'm pretty sure isn't "namaste." She revs the engine and hurries past the stop sign, sparing no glance in either direction. I sense an episode of road rage coming to a theater near you.

Although I'm sure he'd been the first participant in this little stalemate, the gentleman in the white F-150 across from us was seemingly content to bide his time. And who can blame him? He exhibits an air of amused patience that comes with age. I like him already.

A glance in the mirror finds the co-op driver several car lengths back down the slope. I would wager to say the proximity of his combustible fuel to the gray smoke trailing from Gertie's bed played a large part in his decision to reverse course. Caution outweighs the possibility of a big *boom* any day.

Well, you know what they say—go big or go home. So, living up to my fake-it-till-you-make-it personality, I stomp on the accelerator. Gertie surges through the intersection, and we coast through the back entrance of Mueller's fruit orchard.

Passing Mr. F-150, I raise my index finger from the steering wheel. Those who live in rural communities prefer this greeting over the big city wave, which I understand uses an entirely different finger. The man returns the gesture, and a wide grin spreads across his weathered face.

"That was Bill Mueller, Sr. He must have come by to open the gate," Karl says.

"I thought he looked familiar. That was nice of him." One less stop and stall for me.

Karl snorts. "He'll be talking this up, down at the *Casey's* tomorrow morning."

"What? Are you afraid he'll tell your little coffee gang that you let your novice beekeeper drive your truck?"

"No——and are you sure you want to call what you've been doing driving?" Karl asks. "And it's not a gang."

"Oh yeah? Then what do you call a group of farmers who meet for coffee and donuts every morning?" Out of deference to men of Karl's generation, I don't use the pejorative "old." Instead, I begin the laborious process of cranking up the driver's side window. Manually.

"I call 'em good friends," Karl replies.

Nice comeback, Karl. "How come you never invite me to join you?"

"Didn't think you would come," he says. "Figured you only drink those froufrou coffees."

He points at the recycled cardboard cup propped gingerly in my lap. Half of my Venti Soy Latte is splattered across the legs of my white coveralls. Crap! Did I mention the lack of cup holders in this demon——correction: demoness——of a truck?

Karl grunts, a slight smile on his face.

Yikes. I do need to get a life. I'm not even thirty, yet here I am, begging to hang out with a group of senior citizens who drink day-old coffee and talk endlessly about the weather.

I slide my cell phone into the interior pocket of my protective coveralls. Turning in my seat, I notice Karl struggle with his window. And, I wonder, for the thousandth time, why the older man doesn't buy a new truck. Preferably one with automatic everything, minus the attitude. "Shouldn't we leave the windows cracked?"

"Nope," Karl says. He points toward the bank of gray clouds just above the tree line. "Rain moving in——so, close 'em up. We don't want to pick up any hitchhikers."

By hitchhikers, I know Karl is referring to bees. More specifically, the *Apis mellifera carnica*, commonly known as the Carniolan honey bee. The beekeepers love the genus for its calm and docile temperament, so I scan the interior for any bees before closing the window. A trickle of sweat collects in the cleavage of my bra. Gertie does her best to thwart me when I put my shoulder into the heavy door. At last, the door opens, and I practically tumble onto the ground. Behind me, I swear the truck snickers.

Once outside, the sun bounces off my white beekeeping suit, making me grateful for the dark tint of my sunglasses. I shove a plastic pith helmet on my head, pull the folding veil over the brim, and secure it to the top of my collar. The pair of ventilated leather and canvas beekeeping gloves extends nearly to my elbows, and velcro cuffs around my ankles fit snugly over the tops of my boots. "No skin, no stings" is my rationale.

In contrast, Karl wears a simple long-sleeve work shirt with a green mesh hoodie. This Bug Baffler protects the wearer from insect bites and is a lightweight alternative to the traditional beekeeping suit. Karl's short leather gloves remain tucked into the back pocket of his pants. The juxtaposition of master versus novice beekeeper is not lost on me. Karl is a show-and-not-tell kind of guy.

"Stoke up your smoker," Karl says. Pumping the bellows, he coaxes a steady stream of pale gray smoke.

I snatch my smoker and hurry after him, recalling the day Karl taught me to light it properly.

"A good fire needs three things," Karl had said. "Oxygen, fuel, and heat. Do you know how to build a campfire?

"Of course. I was a Venture Scout."

"Venture Scout? What's that? Some woke Girl Scout?"

I'd laughed aloud, remembering how out of touch a man in his eighties can be. "Karl, you must be watching way too much FOX News. In this century, boys and girls camp out together." In truth, my foray as a Venture Scout lasted only until I spent a week camping in Colorado with stinky fifteen-year-old boys and no hot showers. No thanks!

I'm gratified when a puff of gray smoke wafts from the tip of my smoker. Ha—take that, Venture Scouts.

Mueller's, a small family orchard, is located west of Colette, Kansas. Various cold, hardy peaches, plums, and cherries are sold from their roadside stand every spring. And then, in early fall, pumpkins dot the fields beside the apple trees. Carloads of families trek to Mueller's to pick apples, buy pumpkins, and drink fresh apple cider. As a young girl, my family had done the same many times. The memory remains bittersweet.

"Do you hear that?" I call out to Karl's retreating silhouette.

"Bees." He points upward.

The canopy of blooms quivers under the onslaught of worker bees. Their frantic activity is an indication the honey flow is on. I crane my neck to watch.

The mesh veil partially obscures my vision as the first group of beehives comes into view. Earlier this spring, my younger sister and I decorated each wooden brood chamber with a pollinator-friendly flower design. Not only do the flowers add a pop of color to the plain white boxes, but I like to believe they help identify hive temperament by classification. For example, the black-eyed Susan (*Rudbeckia hirta*) is a hive of a mild character. It forms the anchor in this group of five colonies.

Generally, hive entrances face southward. The purpose is twofold: to receive the warm spring air pushing north from the Gulf of Mexico and to help orient young colonies to their new locations.

These five nucleus colonies quickly built out comb on the wax foundation of all twenty frames found in the two deep chambers of the Langstroth hive bodies. And today, we will add the shallow boxes of frames called supers, in which the bees store surplus honey.

Hive tool in hand, Karl removes the brick weights from the top lid of the black-eyed Susan hive. He uses the flat end to pry up the corners of the inner cover, which is sticky with propolis, the glue-like substance bees produce by masticating beeswax, tree resin, and saliva. Honeybees flow lazily from the top of the frames, seemingly unconcerned by this distraction. Instead, they focus on retrieving pollen from the preponderance of blooms in the orchard.

Karl lifts out the first frame to the right side of the top deep brood chamber. Over his shoulder, I notice the beeswax foundation is capped with honey. The center frames show a solid pattern of brood with pollen packed tightly around the edges, all signs of a fertile queen and healthy hive. We move quickly down the remainder of the first group: primrose, coneflower, goldenrod, and hollyhock. Each resulted in ten full brood, pollen, and capped honey frames.

"These look good. Let's go check the other ones," Karl says. "Then we'll return with queen excluders and honey supers."

We head to the second group on the orchard's south side. This grouping has been highly productive as a recipient of the southernmost exposure. Each colony is stacked two deep with honey supers, a few partially filled.

Despite my sunglasses, I'm forced to shade my eyes from the glare. I notice the top of the sunflower (*Helianthus annus*) appears cockeyed. A trick of the light? Perhaps. We press forward, and I realize the top lid and inner cover are missing. The two narrow honey supers lay smashed to the side in a messy heap of broken wood. Confused bees fly erratically from the open top of the brood chamber, then back and forth between the shattered frames.

"What the—" Karl exclaims.

How I failed to notice the pile of white clothing lying crumpled alongside the broken frames is one of life's unexplained mysteries. But, I'm ashamed to admit, I froze for what seemed like an eternity, rooted in place.

Adrenaline floods through my system as the two tiny almond-shaped amygdalas on either side of my brain trigger my instinct to flee.

It's a body! More precisely—the body of a man dressed in a white beekeeping suit, lying perfectly still.

"Karl! The bees—they're all over his face!" I drop my smoker and sprint to the man's side, immediately pressing my fingers into his neck to check for signs of a pulse. Undeterred by my intrusion, a mass of bees crawls about the lower half of his face. I feel no pulse through my thick gloves.

"Your gloves—" Karl says. "Here, use the smoker on the bees. I'll check for a pulse."

Panicked, I begin to pump the bellows. At last, a steady stream of smoke erupts from the tip as the bees scatter from the man's face. I notice a chunk of honeycomb oozing with viscous honey protruding between his lips. His airway is obstructed by beeswax.

Karl kneels to press his bare fingers against the man's carotid artery. "Nope, I don't feel anything. Ouch! They stung me," he says, snatching his hand away.

Two of the knuckles on Karl's left hand turn red and begin to swell. Unlike its cheerful name, the disposition of the sunflower hive tends to be aggressive. The bees buzz angrily around our veiled faces, and Karl shoves his swollen fingers into his gloves. I reapply the smoke, attempting to disperse them. But, once the smoke recedes, the bees speed back to feast on the wad of honey and comb stuck between his teeth.

"We need to clear his airway and start CPR! He's probably been in anaphylactic shock! I'll grab the EpiPen from inside my first aid kit."

Gently, I pry the lump of beeswax from the man's mouth. All the while, Karl applies smoke. Leaning to inspect the man's features, I notice his lips retain the bluish tinge of prolonged cyanosis. After wiping my gloves down the front of my coveralls, I pry the man's eyelid with one finger. The pupil is fixed, and the iris is non-reactive against the glare of the bright sun. Perhaps it's the invasive intimacy of the moment, but something about the man appears vaguely familiar. Still, I fail to recognize him between the haze of smoke and bees. I would have run from the orchard and never looked back if I knew what I didn't know now.

"Show me what to do while you get the kit," Karl says.

Quickly demonstrating the correct hand placement and rhythm for cardiac compressions, I pull off my sticky gloves and fling them aside. The bees speed after them. Deep down in my gut, I know the man is dead, and our actions are futile. But my heart urges me to try. I retrieve the phone from my coveralls and slide my bare fingertip across the screen.

"911, what's your emergency?"

"There is a man down at Mueller's orchard!" By the time I reach Gertie, I'm all but panting the words. "He is covered with bees and doesn't have a pulse!" Grabbing the bright red first aid kit beneath the bench seat, I plead, "Please hurry."

"Did you say bees?" the operator repeats. "Ma'am, please stay on the line."

Karl's veil is pushed back from his face. A line of red welts dots the right side of his cheekbone. "Karl, your face––you've been stung. Again."

"The girls are pretty agitated," Karls says, his breathing between compressions ragged. "Can't say I blame them."

"Can't say I do either, but stop for a second and let me inject the Epi." I jam the auto-injector into the man's upper thigh. "I'm not sure how long before this takes effect. Or— if it will?"

"In all my years of beekeeping, I've never seen anything like this!" Karl says.

"On the way to the truck, I realized I know who he is."

Karl resettles his veil loosely over his head. "You recognize him?"

"At first, I didn't— but, yeah—he's Prof. Robert Fontenoy, the head of the Science Department at Colette College."

"Colette College? What in the world is he doing out here?"

"I have no idea."

In the distance, the wail of sirens resonates. I resume compressions to circulate the epinephrine. Karl trots off to direct the ambulance to our location. Looking into Prof. Fontenoy's vacant stare, dread sweeps over me. To discover a man lying dead among my beehives cannot be good, especially not for poor Prof. Fontenoy nor, I suspect, for me.

Chapter Two

In the periphery of my vision, I notice the flashing lights of a Corley County Sheriff's Department SUV abruptly shut off. The vehicle rolls to a stop far from where I continue chest compressions. The rhythm is on full auto in my head. *One and two and three and four. Five and six and seven and eight.*

Where is the ambulance? My efforts are flagging. My coveralls cling to my body, drenched in sweat. My ragged breathing echoes in my ears. Miraculously, the bees have left. And from my position over the professor's body, I see no visible signs of bee stings. Although I find the absence more than curious, I'm aware that in a case of severe anaphylaxis, one sting is often more than enough.

Nine and ten and eleven and twelve. The passenger door opens, and Karl hurries toward me.

"The new sheriff is concerned about the bees, so I loaned him my gear." Karl picks up his discarded smoker and begins to stoke the bellows. "I'm gonna close up this hive and use the smoke to get the stragglers back in. Any change?"

Thirteen and fourteen and fifteen. "No—I think it's been about ten minutes since we injected the Epi, and there's been no reaction. Where is the ambulance?"

"The ambulance, the fire truck, and at least three other Sheriff's Department vehicles are parked at the back gate. I believe they're rounding up extra bee gear for one of the paramedics who is severely allergic. But the new sheriff is here."

I glance up to see Karl drag the jumble of frames off one side. He picks up the inner cover and slides it atop the remaining hive bodies while vigorously pumping the bellows. "Seriously, Karl—you need to get those stings looked at." *Sixteen, seventeen, eighteen.*

"I will." Karl refocuses the smoke into the center hole of the inner cover board. The bees begin to crawl downward, an evolutionary response I'm grateful for. *Nineteen, twenty, twenty-one.*

"Sounds like good advice, Mr. Bauman. You should listen to the lady. The ambulance is on its way back here. Have the EMTs check your face; it's pretty swollen."

Through the smoke, a tall man approaches with Karl's Bug Baffler stretched tightly across his muscular frame. He waves a hand to clear the smoke and kneels across from me.

"Stop compressions while I check for a pulse," he says.

His voice is low and deep with a pleasant timbre. Through the dark green mesh, he looks briefly at me. I notice a black patch covering his left eye. Complying with his order, I watch him remove a tactical leather glove and place two fingers on Dr. Fontenoy's carotid artery.

"No pulse." He positions himself over the man's sternum and delivers two rescue breaths. "I'll take over from here."

Grateful, I relinquish my place. His interlocked fingers fall into an easy rhythm. *One and two and three.* The man is well-trained, and I recall hearing that our new sheriff is a veteran, the eye patch from an IED explosion in Afghanistan.

"Mr. Bauman tells me you recognize the victim?"

Four, five, and six.

"Yes—I do. I mean, I did—" I stammer, the sudden question and the break in the rhythm taking me by surprise. "The man is Robert Fontenoy, a professor at Colette College—I mean Colette University. Er—I can't quite get used to saying University." Boy, did that sound lame or what?

Seven, eight, and nine

"I'm Interim Sheriff Sean Riordan. Can you state your name and relationship with Robert Fontenoy for the record? No pulse, continue."

Ten, Eleven, twelve, and—

This seemingly innocuous question confuses me. For the record. What record? "My name is Indie—" I stutter. "Ah…I mean, for the record, my name is Indigo Bleu Evans. But most people call me Indie." The words slip out unbidden, and my frayed nerves unravel. "I don't have a relationship with Dr. Fontenoy. I took his freshman Entomology 101 course, but that was four years ago." TMI. Slow your roll, Indie—way too much information.

"So, you say he was a bug doctor? "And your name is Indigo Bleu? That's an unusual name for a beekeeper."

Thirteen, fourteen, fifteen.

"I wouldn't call Prof. Fontenoy just a bug doctor, Sheriff."

"I'm only the acting sheriff until Sheriff Kramer returns from medical leave, so it's just Sean. And, out of curiosity, what would you call Prof. Fontenoy?"

Sixteen, Seventeen, eighteen—

"Well—Sean—I mean—he *was* the chair of the Science Department." And the man with the power to derail my academic career. For obvious reasons, I keep this information to myself. "This is my first year beekeeping." I know the words sound lame as soon as they leave my mouth. "And you wouldn't be the first to say my name is unusual—that's the story of my life."

"Didn't expect to find you out here, DB."

A fireman looms behind Sheriff Riordan's crouched form. But, his childhood taunt betrays his identity as one of the sheriff's deputies wearing a flame-retardant coat and helmet with the shield pulled down over his eyes carrying an extinguisher and a small acetylene torch. Chief Deputy Harlan Pierce is hidden behind the visor. Inwardly, I groan—my childhood nemesis.

Another fireman crouches alongside the sheriff, wearing a heavy jacket emblazoned with the word PARAMEDIC. "Stop compressions." Unzipping the front of Prof. Fontenoy's white coveralls, he places a black stethoscope atop the chest wall and listens carefully. "Nothing. I'll pop him with another Epi and try the AED."

The paramedic runs a pair of surgical scissors down the center of Dr. Fontenoy's green T-shirt and slaps two large AED patches on the right and left sides of the exposed skin.

"You want me to use foam on the bees?" Deputy Pierce asks. "Afraid it's too close to use the torch. It could set off a fire in the trees."

"Fire! Foam! What are you talking about, Harlan?"

"I'm talking about exterminating these killer bees. And that's Chief Deputy to you, DB," Harlan says. "Best if you stand back while I take care of things." He pulls the pin from the handle and aims the nozzle toward the beehives.

"Don't you dare." Jumping to my feet, I remove my veil and pith helmet and fling them at Harlan's chest. They bounce harmlessly off his heavy gear. I push my sweat-soaked hair back from my forehead and storm over to stand in front of my hives, arms wide in a protective stance. "Do not touch my bees, Harlan Pierce!"

"You need to move aside, DB, before I have you arrested for assaulting an officer of the law," he says.

The chief deputy lowers the nozzle and sprays a tiny squirt of foam across the toes of my boots.

"Deputy Pierce put that thing down." Interim Sheriff Riordan's voice snaps whip-like in the air. "I'm a little confused," he says. "I thought your name was Indigo."

"It is, Riordan," Harlan chortles. "It's Indigo Bleu, and both words are the same color—ergo… double blue. So I call her 'DB' for short."

"Okay, okay, that's enough, Pierce. And that's Interim Sheriff Riordan to you, Chief Deputy. Now put the extinguisher and radio for the ambulance. Mr. Bauman seems to have the bees in hand, so it's safe to return. We are not exterminating anything today. Do you understand?"

"But Megan is allergic," Harlan says.

"Charging. Clear," the mechanical voice rings out.

I look away when Prof. Fontenoy's bare chest jumps under the voltage of the AED.

"No rhythm detected," the machine says. "Charging. Clear."

We watch the machine cycle through three revolutions of subsequently higher discharges. Again, Robert Fontenoy's heart does not respond. The potent smell of singed hair mixed with other bodily odors makes me want to gag. The paramedic places his stethoscope on the professor's chest again. He glances up at Sean and shakes his head.

"We should call it." The paramedic looks down at his wristwatch. "TOD, 11:59 AM."

Death strikes high noon at the OK Corral. I feel immediately chastened at the ghoulish notion.

Somehow the bee-phobic paramedic called Megan has procured a fully protective beekeeping suit. She emerges from the back of the ambulance, pushing a collapsible stretcher. Her white coveralls are immaculate, a far cry from my grimy ones.

"It's okay. The bees are cleared away," Sean calls out. "Here, let me help you with that."

"I've got it," Harlan steps forward. He muscles his way between the acting sheriff and the end of the stretcher, which causes him to drop the fire extinguisher. It hits the ground with a thud. A steady stream of foam covers the lower half of my coveralls.

"Idiot."

"Huh? What did you say, DB?" Harlan asks.

"I said you're an—oh, never mind, and stop calling me 'DB.' You remember what happened to you the last time—"

"Hey, Riordan—I mean—Interim Sheriff Riordan, she's threatening me," Harlan whines.

"Enough! Pierce, give me that thing." Sean Riordan grabs the metal canister and reinserts the pin. The white foam slows to a trickle. He shoves the extinguisher toward Harlan. "Let's stay on track, people. Roll him over so we can examine his back. And where is the camera?"

The paramedics comply, and Megan swats her hands before her veiled face. A couple of errant bees fly off the ground near the body. Geez, what a drama queen.

"Careful," Sean warns.

"Looks like we've got a head injury!" One of the paramedics proclaims.

I peer over their heads and instantly wish I had not. Dried blood cakes the large gash visible on the back of Prof. Fontenoy's skull. A dark pool has congealed on the ground beneath the imprint of where his head lay. I feel my stomach churn in earnest.

"Looks like a depressed skull fracture with an open wound approximately three inches in diameter to me." Megan probes the wound with gloved fingers. "And the severity and location do not appear consistent with an accidental fall. I'm only speculating, but it appears as if he was struck from behind. Maybe with one of those bricks weighing down the lids?"

"So, you don't think he could have fallen and accidentally hit his head on the edge of one of these wooden hives?" Sean asks.

"I can't be positive, and we need to wait for the M.E., but this looks deliberate by the angle and coagulation pattern. Maybe as he turned away from the perpetrator? And, besides, if it were an accidental fall, he wouldn't be positioned face-up like this. See these marks." Megan points to the blood-soaked dirt beneath the body. "He was rolled over after he was struck."

"Then the perp must have stuffed his mouth with that honeycomb. Man, that is seriously messed up," Harlan says. "So, you are saying this is a homicide?"

"One poorly staged to appear as a bee attack," Megan adds.

Feeling uneasy, I look up. "What? You can't think I had anything to do with this?"

"Well, they are your bees," Harlan says. " Didn't I hear about you getting kicked out of your graduate program not too long ago? I can imagine you were unhappy with the Science Department at Colette afterward. Doesn't Bob Fontenot head that department?"

Somehow, I've lost all sense of self-preservation in my agitated state. My brain screams, "Shut up," but my lips refuse to comply. "What? How—how did you know about that, Harlan?" Okay, if you call trashing your master's thesis over allegations of plagiarism a mere dust-up, then yes. Maybe. But I'm no killer."

"It's a small town. I keep my ear to the ground," Harlan says. He taps the side of his helmet with a gloved hand. "Sounds like motivation to me."

"I—I didn't have anything to do with this," I insist. "This is insane. Why would I? For Pete's sake, I tried to save him. Ask Karl!" I look around for the older man and see him nudge an object on the ground next to the open hive with the toe of his boot.

"Over here," he calls out. "I've got a top lid brick. It's got some blood on it."

"Don't touch that!" Harlan stalks toward the older man. He bends down and flips up the visor on his face shield. "And now we have means."

Harlan stands upright to face me.

"Only question is—when did you have the opportunity, D—I mean Indigo?" Harlan asks.

"That's quite enough speculation, Chief Deputy," Sean Riordan warns.

I'd failed to notice he'd come up beside me, radio in hand. The squawk of the shoulder mic makes me jump.

"Roger that," he says. "ETA on the M.E. is less than ten." Sean Riordan speaks loudly over the radio's static. "Let's cordon off the area and let Tyson finish photographing the scene. We need to establish a grid. Rain's in the forecast for tonight, so let's get a move on, people."

Sean touches my shoulder and says in a low voice, close to my ear, "Ms. Evans, perhaps we could talk over by my unit. I have some additional questions to ask you."

"You want me to cuff her?" Harlan asks.

At least my age and good sense overcome my compulsion to kick Chief Deputy Harlan Pierce in the nuts. Instead, I dutifully follow the new Interim Sheriff to his SUV.

Chapter Three

- -

"**W**ait a minute—the sheriff let you leave? Just like that?" Heather asks. After a shower and change of clothing, I'd snuck down the back stairwell of the loft apartment I shared with my sister on the third floor of The Hive, a cavernous brick building we leased and turned into an artist co-op.

Every beehive has a queen. Our retail space is appropriately called the Bee Queen. We specialize in all things honey and carry both imported and local varieties. In addition, we sell honey-infused lotions, soaps, and balms produced on-site in our newly approved commercial kitchen. The wall space is adorned with honeybee art, jewelry, and pottery created by artists within the co-op.

"No, Sean, ah—Interim Sheriff Riordan, let me come home to clean up. He said someone would be over to conduct a more thorough interview later. They wanted to finish processing the site before rain moved in. Although, Harlan Pierce about threw a fit. He started calling me 'DB' again. I swear he was ready to lock me up."

Always one to pick up on the slightest nuance, Heather queries, "Oh, Sean, is it? I've seen the new sheriff around town. He's pretty hot in a toxic masculinity sort of way. And ignore Harlan Pierce. He never got over you nailing him with that dodgeball. When was that? Like—junior high?"

"It was seventh grade."

Heather reaches beneath the wood countertop and pulls out a white bakery box. "Here, try one of these honey-lavender scones from Angela's counter. You'll feel better with something in your stomach. Also, I've brewed a pot of Himalayan white tea from Chai. Let me pour you a cup?"

"No! I don't want a cup of tea!" Seriously? My sister never misses a chance to advertise our vendors' wares. Even in the face of impending disaster. "Did you not hear what I said? Dr. Robert Fontenoy—the same Dr. Fontenoy who all but accused me of plagiarism was found dead next to my beehives. The new sheriff might be hot, but I'm sure he thinks I'm somehow involved!"

"Are you sure? The tea complements the scone." Heather places two hand-thrown pottery mugs atop the counter. "And you do think he is attractive. I knew it."

"No, it's not like that. I—" My stomach begins to growl, and I realize I haven't eaten a thing all morning, so I take the proffered scone and bite it. It is delicious. "I should call Dad. He's a lawyer. He'll know what to do."

"But still, "Heather continues, "you had no reason to want Prof. Fontenoy dead. If you wanted to kill someone, you would have killed Alex."

I hadn't spoken to anyone outside of Heather about my breakup with Alex Carmichael. The wound was still too raw.

"Who do you want to kill?"

Ugh. Our mother sweeps into the boutique. And by sweep—well, suffice it to say, Claire Evans, MFA in Fine Art and Fiber Studies from Colette University, is a woman who knows how to make an entrance. Claire looks chic in a fitted white tee and tapered silk trousers topped by one of her hand-woven scarves. The loose weave of teal and green tones derived from plant-based dyes accentuates her green eyes and deep red hair.

I cup my hand to my ear. "Perhaps you should speak up. I'm sure only half of the downtown residents heard you by now." And if I had to bet, the other half would know soon enough.

"Mother! You look lovely today." Heather holds up the bakery box. "You should try one of Angela's scones?"

"Not this morning, dearest. I'm on a cleansing fast. And who else would you want to kill if it's not that cheating ex of yours?" Claire asks.

17

"Could you at least try and be discrete, Claire?" Annoyed by her cheerful tone, I deliberately use her first name, knowing it gets under her skin. Her skin does look exceptionally luminous, with nary a fine line or wrinkle to be seen. "Are you using fillers again?"

"Indie," Heather warns.

Petulantly, I ignore my sister. My mother tends to bring out my worst instincts. "Is there something different about your hair? Did you color it recently?"

"Fine, Indigo," Claire snaps, touching her copper curls. "If you prefer once again to shut me out, if you prefer to treat your mother like a total stranger, then so be it."

"Indie—" Heather pleads. "Tell her."

"Alright, alright. Tragically, Dr. Robert Fontenoy was found dead beside my beehives in Mueller's Orchard this morning. I'm afraid the Sheriff's Department thinks I might have had something to do with it."

"Fontenoy!" Claire exclaims. "Bob Fontenoy from the Science Department? Dead? Well, that can't be. I just spoke with him at the summer faculty mixer Tuesday evening."

"You talked with Dr. Fontenoy?" I ask.

"Wha— wait. You went to a Colette faculty mixer?" Heather asks. "Why?"

"Of course I did," Claire snaps. "I am an adjunct faculty member, and Bob Fontenoy is, or was, faculty. But I can't believe the poor man is dead."

"Well, he is—" The words feel no less surreal when I speak them aloud. "But Heather's right. Since when do you care about mingling with other faculty members, especially someone in the Science Department?"

"Oh, for heaven's sake, I barely spoke to the man. He approached me and rambled on and on about some new research grant the department qualified for. You know, I detest anything to do with GMOs and pesticides," Claire says.

"Pesticides?" Claire's words reverberate in my head. "Are you talking about the grant awarded by Agro-Tech Industries?"

"I have no idea, Indie. I confess to tuning out of the man's conversation after we discussed you. In retrospect, I probably should have paid more attention." Claire shudders. "Oh, poor Bob. Who could have imagined he would be dead two days later."

"You spoke about me with Prof. Fontenoy? Thanks for burying the lead, Mother. What did he have to say?"

"Hmm…" Claire pauses. "Oh yes, I remember—he said you were a fine student with a bright future in your field of study. Yadda, yadda, yadda. You know, the usual chitchat amongst colleagues."

"And how did you respond to him, Mother?" Heather asks.

"Well, I agreed, of course. I told him it was a shame Indie decided to throw away her academic career over nothing more than a misunderstanding."

"A misunderstanding?" I feel my face begin to flush. "The man accused me of using non-original data in my thesis. I can't believe you would reduce plagiarism accusations into something so—prosaic. He rejected my premise in an email based on—what? And then, when I challenged him, he referred me to Professor Carmichael. As if I would ever speak to Alex again after what he did!"

"Indie," Heather soothes.

"Well, I wouldn't know what Alex did. Would I?" Claire snaps. "Since you've refused to tell me why the two of you broke up. Although, I have my suspicions. The man was much too good-looking to be a scientist." A chime sounds. Claire glances down at her smartwatch, tapping the alarm. "Sorry, girls, I am off to Kansas City to pick up our new artist-in-residence. He arrives from Italy today, and as my colleague, Dr. Peavis, is off on sabbatical, I've been tasked with helping him settle in."

"From Italy? Wow, so exciting. I had no idea. That's kind of you, Mother." Heather says.

"Yeah," I agree. "It is kind. What's the catch?"

"I assure you, Indie, there is no catch," Claire insists. "Why must you always be so suspicious?"

I examine my mother's appearance closely. Newly coiffed hair, expertly applied makeup, and a fresh manicure rather than fingers stained from the plant dyes she mixes for weaving yarn. Dyes she makes from natural pigments like the indigo plant. Okay, I might have a few unresolved mommy issues. But how would you feel being named after a blue dye? Still, my sister was spot-on, and I wasn't wrong to be suspicious. Claire never socialized with other faculty members or did favors for Dr. Peavis, whom she despises. Something was up.

"Oh, hello, dears. I hope I'm not interrupting anything?"

At the sound of a newcomer's voice, we turn to find a petite, older woman dressed entirely in hot pink. I have no idea how she managed to sneak up on us clad in such a flashy color. The color scheme flowed from her pink-tinted hair down to the toes of coral pink Sketchers. Even though the woman's pink tunic sported the words "High Priestess," spelled out in opalescent pink sequins and worn atop pink capris. I feel my eyeballs cramp looking at this senior citizen version of the PBS character Pinkalicious, I feel my eyeballs cramp up. I glance at my mother and sister in turn. Each wears an equally stunned expression across her face.

"Bea!" Claire reaches to embrace the tiny woman. "I felt empowered by your reading session yesterday. I can't thank you enough."

"Wow, Granny Bea, you're very ah— pink today," Heather says. "It's good to see you. I hope Wyatt was able to help you settle in?"

Okay, so who is this tiny, pink apparition, and why am I the only one who doesn't seem to know?

"Oh yes, dear. Wyatt is such a nice young man. He's set up my social media pages. Now, Sage Wisdom is officially on the interweb," Bea says.

Sage Wisdom? Interweb? What is happening? I watch the woman slip on a pair of hot pink, cat-eyed readers. She squints at me from above the rims and holds out her hands.

"You must be Indigo. I've heard so much about you," Bea says.

Obligingly, I place my own within hers and watch as the tiny woman turns them over and begins to study my palms.

"Yes—" Bea says. "I see your lifelines are very intriguing, Indigo. Have you ever noticed how your career line ends here just as it connects with your love lines? A sign you've been disappointed. Perhaps in one or possibly both?"

Bea closes her eyes and begins to sway back and forth. "Ah... ohm..." Bea hums. The soft sound gurgles in her throat. "I sense your root chakra is quite murky, my dear. Is there something troubling you?"

Aghast, I look to my sister for aid. "What the—?" But Heather shrugs her shoulders and smiles sweetly at me. Thanks for nothing, baby sister.

"Why, Bea, however, did you know? Indigo is under investigation for murder," Claire exclaims.

I snatch my hands from Granny Bea's and level Claire and Heather with a blistering glare. "Is this a joke? I am not under investigation." Not that I need to justify my innocence to this pink harbinger of doom. "I don't know you, but eavesdropping is a serious violation of privacy."

Heather shakes her head, and for once, my mother is speechless. The little woman continues to sway until Heather gently touches her arm. Bea's eyes pop open, and she looks straight up at me.

"Danger—nothing is as it seems," Bea whispers.

"Ah well, girls, I must be going," Claire says uneasily. "The traffic into Kansas City can be busy this time of day." She pats Bea companionably on the shoulder. "We'll talk soon, Bea. Oh, by the way, Indie, I remember another tidbit of conversation with Bob Fontenoy. And no, he did not give me a reason before you ask. But he told me he hoped to have an opportunity to chat with you. Although, I guess now it doesn't matter."

And just like that, my mother, never one for confrontation, hurries out. The three of us stare silently at her retreating backside.

"Oh, Indie, you haven't met our new tenant, Beatrice Kowalski or, as she prefers to be called, Granny Bea. Her business, Sage Wisdom, has opened in space twenty-three. She is a tarot reader and spiritual advisor," Heather says.

"Oh yes, dears, it's my lucky number," Bea interjects. "You see—two plus three equals five, which is my life path number. So, I suppose it was meant to be. But, cross my heart, Indie, I would never listen in on a private conversation." Bea intersects the sequined letters on the front of her chest with her fingertips. "I would love to offer you a free reading at your chosen time."

A reading. That would be a big fat NO! I am hardly a devotee of woo. All the while, my mother's words replay continuously in my head. Why had Prof. Fontenoy wanted to speak with me? It must have had something to do with my thesis. I look down at the pink wannabe seer. "While I appreciate the offer, I don't believe anyone can predict the future, so no thanks. But welcome to The Hive., I guess?"

Bea chuckles. "Oh, my dear, I see you are not a believer. But I predict perhaps someday you will be."

"I highly doubt it." Never in a million years.

"What if I were to say you are about to get a visit from a tall, dark, and handsome stranger?" Bea asks.

"I'd say you need to come up with something more original."

"Fair," Bea acknowledges.

"Besides, I find tall, dark, and handsome highly overrated."

Heather clears her throat. "*Ontday ooklay ownay*," she says, cupping her mouth with one hand. She rolls her eyes toward Angela's Eats and Sweets kiosk.

"What did you say?" Why is my sister resorting to our childhood code in front of a stranger? Instinctively, I turn.

"It's pig Latin, dear," Bea says kindly. "I believe it means 'don't look now.'"

Of course, I looked, and there he was, Interim Sheriff Tall, Dark, and Handsome, striding our way. Heather was right. The man positively reeked of toxic masculinity.

"How did she know he was coming here?" Heather's voice quivers.

I was about to ask the same question. I whirl back around to find our resident pink purveyor of fortune-telling has conveniently disappeared.

Chapter Four

"**M**s. Evans, may I have a word?"

Sean Riordan disturbs our decidedly feminine space with a w*hoosh* of masculine energy. Seeing his stern face makes me feel like the little mouse that Mustard, our store cat, cornered last week: trapped with no way out. "Do not leave me, Heather," I whisper to my sister.

"Interim Sheriff Riordan." Since he's called me Ms. Evans, I feel the visit deserves the use of his formal title. "Twice in one day. What do I owe the pleasure?" The silence lingers long enough to intimidate. "I believe you've met my sister, Heather Grace. Could I—interest you in some of our local honey products?" I try and use deflection, a tactic best wielded by my mother.

"Now," Sean insists.

The deflection was an epic fail. "Do I need a lawyer?"

"I don't know. I guess that depends on whether or not you have something to hide," Sean says.

"How about I brew a pot of tea," Heather interjects. "The tables in front of Angela's are empty. Indie, take the mugs, and I'll bring the tea right out."

Heather thrusts the empty cups into my hand, and I watch her carry the teapot through the kitchen door, leaving me alone with Sean Riordan. Traitor.

"Shall we?"

"Lead the way."

We settle into a corner table. I notice Sheriff Riordan picks the seat facing the doorway. I guess it is a cop thing. He briefly scans the interior, and the brilliant blue of his right eye contrasts strikingly with the black patch covering the left. Sean is reminiscent of a buccaneer rather than a law-enforcement agent, along with his dark hair and the emergence of a five-o'clock shadow. Not that I would ever find the image of Sean Riordan as a swashbuckling pirate remotely attractive. No—not at all. Aargh.

"Is it always this quiet in here?" Sean asks.

I bristle at the insinuation we lack customers. But, in truth, we do. Similar to many college towns of its size, Colette experiences a significant population decrease in the summer months.

Colette College was founded in 1886 by Franciscan monks and became a fully accredited university five years ago. The university's namesake, St. Colette, was the abbess of the Order of the Poor Clares. The order is renowned for its devotion to perpetual fasting and rejection of worldly income. Similar to traits found in the life of the modern-day college student. The root cause of poverty for most undergrads is their propensity to party rather than devotion to God.

"Unless you are from here or have summer classes or a job, only about a quarter of Colette students stay in town."

"That puts a big dent in my suspect pool," Sean says.

Sheriff Riordan swivels his head toward me, and I feel the flush of prickly heat sweep up my neck. Damn, my red-haired genetics. "Wait—this morning you said I wasn't a suspect."

"And you weren't. This morning," Sean says. "Would you like to amend your statement?"

"My—my statement?"

"You know when you said you 'barely knew' Robert Fontenoy? It seems you left out a few inconvenient facts."

"Indie! Don't say another word without legal representation," Heather advises.

My sister appears next to our table, holding a cell phone in one hand and a steaming blue and white china teapot in the other. A thin young man with dirty blonde hair stands beside her. He carries a plate piled high with leftover cookies from Angela's kiosk.

"Ms.—ah—Heather Grace, is it?" Awkwardly, Sean Riordan pushes back his chair to stand. He reaches for the teapot. "So kind of you to—"

"You can save your charm for someone else, Sheriff," Heather says. "I won't allow you to intimidate my sister, no matter how good-looking she finds you. Here, have a cookie."

She shoves the plate unceremoniously toward him.

"Wyatt, meet Interim Sheriff Sean Riordan of the Corley County Sheriff's Department. Interim Sheriff Riordan, meet Wyatt T. Price. Who is, as of five minutes ago," Heather says, tapping the cellphone screen, "my sister's legal representative."

Sean accepts the cookie and obediently bites into it. "You think I'm good-looking?" His sideways glance holds the barest hint of a smile.

"I never said that." Not precisely. "And what? How can Wyatt be my legal representative? He hasn't even graduated from law school."

Sean pops the remainder of the cookie into his mouth and chews. "Crunchy. What's in these things?"

"They're vegan," My sister and I say in unison.

"And gluten-free," I added smugly. At this time of day, Angela's vegan, gluten-free cookies tend to become extra crunchy and are usually the only treats left in the display case. Secretly, I hope he chokes on the chickpeas and almond paste. Alas, I should have known a man accustomed to MREs on the battlefield could eat anything.

"Hmm… not too bad." Sean brushes the crumbs from his lips. "Now, Mr. Price, is it? I don't think you can represent Ms. Evans if you are not an actual lawyer." He modulates his tone. "We can make this official and question you at the station if you prefer."

"Actually—" Wyatt says, pushing a pair of black-framed glasses up on the bridge of his nose. "I've just completed my second year of law school, and under Kansas state law, as a third year, I'm permitted to represent Indie while under the supervision of a licensed attorney."

"Yeah, and who might that be?" Sean challenges.

"Professor Daniel Evans, J.S.D," I interrupt before Wyatt can speak. "My father."

"Who is awaiting our call if you care to verify." Heather waggles her phone. "Should I call him?"

"Huh, well, you learn something new every day," Sean Riordan says. "I've got an idea. How about the three of us sit and have a nice cup of tea? Maybe another cookie? I want to go over your statement. I'm particularly interested in your prior interactions with the deceased. Just you, me and now, I guess, your—legal representative?"

"What about me? I'm her sister," Heather asks.

"Heather," I warn.

Sean raises his palms in the manner of a supplicant. "My apologies, Ms. Heather Grace. And might I say, you've certainly lived up to your name. You've been extremely gracious." Sean smiles and indicates the tea and cookies. "But, since Indigo has obtained counsel, we should probably abide by the tenets of attorney-client privilege. Don't you agree, Mr. Price?"

Somehow Wyatt has acquired a yellow legal pad. He tugs on his sparse, reddish-blonde soul patch, and his prominent Adam's apple bobs up and down in his neck. Does the boy ever eat?

Heather narrows her eyes and plops the teapot down on the table. "Very well, but I'll be right over there." She points toward the arched doorway dividing the Bee Queen from the common area. "Don't think about railroading my sister into some kind of confession."

"Yes, ma'am," Sean Riordan replies with an amused grin. He watches Heather disappear through the archway and settles back into his chair. "I'll say one thing about her—your sister's got game." A note of admiration fills his voice. "Shall we proceed? Tea, anyone?"

"That she does," Wyatt agrees, his gaze lingering on the doorway.

"She's practically engaged!" Instantly, I regret my childish outburst. Oddly enough, I feel slighted. For what reason? Being represented by a second-year law student in a potential homicide might have something to do with it. Or perhaps because Sheriff Riordan found my sister gracious, whereas I had been merely unusual earlier? I hate to admit it, but I suspect the latter.

"You said in your statement that you took Entomology 101 from Professor Fontenoy. Is that correct?" Sean Riordan asks.

"Objection," Wyatt says.

"Yes," I say, looking at Wyatt in confusion. "What? It's the truth!"

"Mr. Price, we are not in a courtroom," Sean chides. "I'm trying to clarify a few of Indie's earlier statements."

Wyatt stares at Sean; a slow flush suffuses his pale skin. "I'm trying to protect my client," Wyatt says. "The question could be misconstrued as entrapment, a way of getting her to perjure herself."

Sean shakes his head in exasperation. "Surely your client can answer the questions without you breaking into legalese every few seconds. So, Ms. Evans, I'm going to ask you once again, have you had any further contact with Professor Fontenoy since your first class with him? It's a simple question. Yes or no?"

"Objection," Wyatt says again. "Badgering the witness."

"This is not a courtroom, and I am not badgering anyone," Sean bangs his fist on the table. His teacup rattles in the saucer as hot liquid sloshes over the rim.

Before Wyatt can raise his hand to object, I interrupt. "Yes and no." Noticing Wyatt and Sean's perplexed looks, I take a moment to compose my answer.

"Would you care to clarify that?" Sean asks. He soaks up the spill with a paper napkin. "For the record."

"Yes, I took Entomology 101 with Professor Fontenoy the summer before I left to attend the University of Wisconsin-Madison. UW is one of the best entomology schools in the country, and I'd gotten an academic scholarship. I believe Professor Fontenoy was appointed Science Department chair the following year. When I came home, Dr. Fontenoy no longer taught classes. Occasionally, I would pass him on campus and say hello. It's a small place, and everyone knows each other. But, other than interdepartmental emails, there has been no contact. Do emails count? If so, I need to change my answer to yes."

"Finally. We are getting somewhere," Sean flips open a padded black folder exposing a slim sheaf of papers. "Care to explain these?"

Wyatt reaches for the folder, but I've scanned the subject line of the top sheet. It's a hard copy of an email I recognize. "I have nothing to hide. I know exactly what each one says."

Sean settles into his chair, eyebrows raised expectantly.

"As I told you, I did my undergrad in Wisconsin and moved back home to pursue a graduate degree." And to follow a man I thought was to be my life partner, but I see no need to bring up that information. My feelings are much too raw. "There should be five in total. In addition, my responses would make—ten?" I direct the question to Sean.

"Go on," Sean encourages.

At this point, I'm fully aware I'm stalling. "The first email from Professor Fontenoy in his position as de facto supervisor of my thesis is a timeline for submitting my prospectus to the graduate committee. The second assigns Associate Professor Alex Carmichael as my direct lab supervisor. The third one acknowledges a committee review is underway. The fourth informs me the committee has rejected my proposal based on evidence of non-original data. The fifth grants a ninety-day extension to resubmit; failure will terminate my program. It's dated—" I look down at my phone screen. "—ninety-two days ago. As you can see, I've missed the deadline."

"Impressive," Sean replies.

"I have a good memory." How could I forget? I'd read those emails ad nauseam, so their contents were etched into my brain.

"Do you have anything else to add?" Sean asks.

Wyatt clears his throat and starts to speak, but I raise my hand to stop him. "No, it's okay, Wyatt. I have nothing to hide." I return Sean Riordan's withering gaze with one of my own. "I would like to state, for the record, I strongly deny the allegations regarding my thesis. I did the work. But maybe I implied in my last email that the committee could kiss my nether regions if they thought I was stupid enough to plagiarize. Someone might have perceived my poor word choice as a threat. However, I take the accusation of plagiarism seriously. A complaint like that can completely derail someone's academic career. Once word gets around, your scientific reputation is toast. Academia is a virtual hotbed of gossip."

"Pardon me, but I wouldn't call the academic world of tweed jackets and scientific tomes a hotbed of anything," Sean says, his tone rich with sarcasm.

"Well, you'd be surprised, Interim Sheriff Riordan."

Wyatt snickers and even Sean is forced to smile.

"Yes, I probably would," Sean concedes. "So, Indigo Evans, can you tell me one good reason I shouldn't arrest you for the murder of Professor Robert Fontenoy?"

"Because I'm innocent. I did not plagiarize the data, and I didn't kill Professor Fontenoy. I don't know who did, but it wasn't me."

"Okay," Sean says. "Just say, for now, I believe you. But if I find out you've been withholding any information applicable to this investigation, I will arrest you for obstruction."

"Objection," Wyatt says.

"Overruled," Sean snaps.

"Thank you." A feeling of relief sweeps through me, and I can breathe again. Now is the time to come clean. "There is one more thing, and you'll probably need to question Associate Professor Carmichael."

Sean sighs and reopens the folder. He flips through the pages and pulls out the second one in the stack. "Associate Professor Alexander Carmichael, B.S. and M.S. in biology and a Ph.D. in environmental studies. I see he was assigned as the proctor to oversee your thesis. What do I need to know about him?"

"I have no tangible proof, mind you, but I'm positive Alex used my study on the effects of neonicotinoids on bee colony behavior as part of a million-dollar grant awarded by Agro-Tech Industries to the Science Department. As you are probably aware, research grant funding is highly competitive, especially at small universities like Colette. Alex's supposed research was integral to the grant approval process."

"A million-dollar grant?" Sean whistles aloud. "That's one way to punch a hole in your academic career ticket. Pesticides, huh? And what did this supposed research on neonicotinoids prove? "

"It was researching alternative and best farming practices to improve the quality of our native pollinator habitats. You know, simple things, like planting vegetation and tree lines alongside crop fields. Believe it or not, such practices provide an inexpensive buffer against pesticide residue and overspray. I'm realistic enough to understand GMO crops are not going anywhere. Especially in the middle of farm country, I feel it's best to work within the industry to create a long-term sustainability plan. Reimagine farming, so to speak."

"How exactly did you obtain this data?" Sean asks.

"Through a controlled study exposing a honeybee colony to graduated doses of commonly used neonicotinoids, then measuring their response."

"Ugh, that sounds like a terrible idea," Wyatt inserts.

"Believe me, and I'm not proud of it. But I stopped the exposure once I got to 100 parts per billion."

"Why? What happened to the bees at that concentration?" Sean asks.

"While the colony didn't collapse, the worker bee activity diminished significantly, and the queen all but quit laying eggs. I reported my concerns to Professor Carmichael, who encouraged me to upload my data to the departmental database. And, like a fool, I did. Incidentally, I'd performed a blind field study testing for residue on milkweed and other pollinator-friendly vegetation. The findings showed the levels of clothianidin were well below 70 ppb. And, believe it or not, these levels are acceptable by the EPA for food and water consumption."

"What exactly is clothianidin—it sounds hazardous?" Wyatt asks.

"Clothianidin is a common pesticide used to coat seeds. It affects the central nervous system of crop pests."

"This is all very interesting, but what does this have to do with the death of Robert Fontenoy?" Sean interrupts.

"Maybe nothing," I say. "But somebody used my data in the grant proposal submitted by the department to Agro-Tech Industries. The study advocated safely using their proprietary weed-killer and soil-conditioner formula, ATI-EX50/50, marketed under Glo-Grow. The levels of *clothianidin* in the product fell well within the acceptable range of 100 ppb. But, as the primary researcher, I felt that long-term systemic effects on honeybees were detrimental. I would never credit Glo-Grow in such a positive manner."

"And you don't think this was a coincidence? I mean, you did say you uploaded your data?" Sean says.

"No, I don't. I carefully secured my findings away from those of the grant proposal's data." In frustration, I shake my head. "Including my data skewed the results verifying the EPA's acceptable standards. But, deliberately omitting ethical considerations seriously flaws the study."

"Okay––so let me guess, the author of this grant proposal was––?" Sean queries.

"Dr. Alex Carmichael. My graduate thesis proctor."

"Of course," Sean sighs, picking up the folder. "A return visit to Colette University is next on my agenda. Hopefully, Professor Carmichael will prove cooperative."

"I should probably mention one more thing." A nervous laugh escapes my lips. "When you speak with Professor Carmichael, don't bring up my name."

"And why is that?" Sean asks.

"Because I sent the fifth email the day I caught him with his teaching assistant. They were…um… together if you catch my drift."

"Yeah, I think I do. Well, I guess that is an unexpected complication. Academia is more of a hotbed than I believed," Sean says, his face completely devoid of emotion.

"Sounds more like a hot mess," Wyatt says.

A hot mess, indeed.

Chapter Five

- -

*B*eep, beep—clank—boom! The sounds cause me to bolt upright in bed. My heart pounds the staccato rhythm of someone dragged from REM sleep. When I hear a metal container settle back into place, I fall back against the pillows. It's garbage day.

Between distant rumbles of thunder and lightning show that only a Midwest storm can produce, further thoughts of sleep vanish. I close my eyes in a vain attempt to recapture last night's dream. A dream in which a certain blonde superhero lays aside his mighty hammer and spoons up against my backside—I wish. Instead, I'd dreamed about bees swarming from the mouth of a faceless man. They'd swooped and dived, then merged into a dark visage strongly resembling that of Sean Riordan, a mass of bees covering one eye.

My bedding may be damp with sweat from a hot body beside me. But it's no blonde superhero. Instead, a fat, long-haired Tabby kneads my side.

"Meow." Mustard, fourteen pounds of hair and flab, glares up at me in the semi-darkness. His gold eyes are wide and unblinking. "Meow," he cries again, and I feel his sides start to heave.

"Scat." I push him none too gently off the bed. He lands with a thud and continues to heave. *Gack!* An enormous hairball falls onto the rug. I hang my head over the bedside to stare down at him. "Seriously, dude?"

Rolling over, I tap my phone screen. Five-thirty a.m. Mustard pads through the open doorway, searching for more peaceful accommodations.

What to do at this hour when you are too afraid to dream? I could get up and go for a run. The air should be relatively cool after last night's storm, but I don't fancy traversing the quiet streets with a killer on the loose.

The wide plank floorboards are smooth beneath my bare feet as I pad down the hall into the open living space. I should adopt a dog. At least a canine companion would provide a modicum of protection, unlike the feckless Mustard. But, between the cat and Lettuce, the Giant Angora bunny who shares our space, we do more than our share of vacuuming. I'm not saying our mother is Cruella de Vil, but as a natural fiber artist, she does have an unhealthy affinity for exotic, furry creatures. I'm loath to admit it's why I chopped off my long hair last year. Sigh—at this point, I should likely seek counseling. Self-consciously, I tuck the chin-length strands behind my ears and wander into the kitchen, flipping on the light.

"Morning, Lettuce." I bend down to release the latch on the wire hutch. A whiskered nose pokes from his den, sniffing the air cautiously. "Yeah, I know it's early, buddy. But I can't sleep, so how about I make coffee or, in your case, a bowl of carrots."

Lettuce hops across the kitchen and stretches up on his hind legs. Fully extended, his reach is thirty inches. His twelve-pound weight is more than a match for Mustard, who has learned to steer clear of the rabbit's powerful hind legs. Lettuce accepts the proffered carrot, and I scratch behind his long, silky ears waiting for the coffee maker to perform its magic.

The laptop on the kitchen table beckons me, and I open the browser and quickly type "Sean Riordan, US Army." Now, I know what you are thinking, but along with a jolt of caffeine, there is nothing better than a healthy dose of cyberstalking to jump-start your day. Coffee mug in hand, I pull up a chair and scour the internet.

Bingo. The most recent article from the *Colette Caller* features a photo of Sean Riordan dressed in a khaki uniform shirt and a wide-brimmed Stetson hat. One hand raised, the other rests on a Bible. Sheriff Kramer stands awkwardly off to one side, a set of aluminum crutches under his arms, if I recall, an early casualty of summer league softball. The blonde holding the Bible is none other than yesterday's impatient driver, Brenda Matthews. Brenda is the wife of our recently elected mayor, Frank Matthews. Why is Frank not in the photo? It's common knowledge our mayor is a publicity hound with political aspirations for higher office within the state. The couple owns Matthew's Real Estate agency, and like any college town, it is a booming rental management business.

Colette's population blooms by two thousand students and faculty during the spring and fall semesters. Colette University retains its cadre of campus security who work in conjunction with the Sheriff's Department. I'm reasonably sure a homicide investigation falls well outside campus security's purview, especially when the case involves one of their own. Law enforcement falls under the jurisdiction of the Corley County Sheriff's Department, and Colette, as the largest town in our decidedly rural county, houses the department's base of operations.

Once again, I refresh Sean's name and type "IED, Afghanistan" into the search window. A headline in Nebraska's *Elkhorn Gazette* rewards me: "Hero's Welcome for Local Son." The news piece summarizes Chief Warrant Officer (ret.) Sean Riordan's medical retirement after suffering an injury while on assignment in the Helmand Province of Afghanistan. Details of the mission are scant to nonexistent, and my curiosity is peaked.

The accompanying photo is of Sean, stern-faced in a full-dress uniform and familiar eye patch, standing alongside an older couple. I scan the article further, and the pair is identified as Marshal Riordan, Sheriff (ret.) of Douglas County, Nebraska, and his wife, Delia. Marshal is a tall, handsome man with a thick head of salt-and-pepper hair. The resemblance between father and son is striking.

In contrast, Sean's mother is petite and slender, with hair more silver than blonde. She's casually dressed in slacks and a knit sweater and has the weathered features of someone who spends time outdoors. Delia's hand clutches Sean's elbow in a gesture of maternal support.

A small group of men and women surround them. All wear the distinctive embroidered hats of the VFW Post 2503, as does Marshall Riordan. So, Sean comes from a family of law enforcement and military service? As a member of the military's Criminal Investigative Division, he is undoubtedly qualified to run a murder investigation.

34

"What are you doing up so early?"

I look up to see Heather walk into the kitchen. She immediately grabs the tea kettle and fills it from the sink. Instinctively, I close the laptop.

"I can't say I blame you. What—between the thunderstorm and thoughts of you finding a dead body yesterday, I couldn't sleep either." Heather yawns and stretches her arms over her head. "Besides, I was up late talking with Austin. He wanted me to tell you he hopes you will be cleared of all suspicion soon."

"I am not under suspicion." Ouch, that hurt, but Heather's boyfriend, Austin Vale, always the do-gooder, sometimes comes up with perfectly useless inanities.

"Of course, you're not," Heather soothes. "He's just concerned about you."

"Concerned? Perhaps, he thinks people will stop buying products from us. Especially if they believe a murder suspect is selling them? I hope you told him I'm innocent. You did tell him that, didn't you?"

"Of course, I told him," Heather says. "But Austin thought we might experience a drop in revenue for a while. You know how people in a small town like to talk."

"I know how people like to make snap judgments." My words sound petty, even to me. "With morbid curiosity and all, we'll probably see an uptick in business. Although, I'd wager Austin is hoping for a drop in sales, so you'll finally run out of excuses to postpone your wedding." Instantly, I'm contrite. "I'm sorry, Heather. I shouldn't have said that." Austin might be off training to be the military's next Captain America, but I know he loves my sister, and I don't want him to think too badly of me.

"Technically, we haven't set a date," Heather says. "And you know Austin always supports my dreams."

"Technically, you keep your engagement ring in your underwear drawer. And you and I know nothing would make Austin happier if you gave up this place and married. Just saying—" I lift the coffee cup to my lips and take another sip.

The kettle starts to whistle, and Heather carefully measures a precise amount of leaves into a mesh tea ball. She pours hot water into a mug. "So, a solution occurred to me last night."

"Oh yeah, and what is that?"

"We should solve the murder," Heather says.

"What—?" I practically spit the hot liquid out of my nose.

Heather settles into a chair, steaming mug in hand. "Think about it. It's a good idea. Not only do we bring Professor Fontenoy's murderer to justice, but we clear your name. I'm positive we can figure this out with your analytical brain and my network of contacts. I wager we can do so before Harlan Pierce and the other Keystone cops at the Sheriff's Department can. So, it's a win-win."

"I wouldn't call Sean Riordan a Keystone cop."

"True," Heather agrees with a shrug. "But the interim sheriff is new to the area. He doesn't know everyone as we do. At the very least, we could help narrow the suspect list."

"What list? So far, I'm the only one on it. Well, I've managed to implicate Alex. Ugh—I feel guilty about throwing him under the bus like that. Appropriating research hardly equates to murder, right?"

"How Alex used his research to enhance his academic bona fides is icky. Then he cheated with his lab assistant—also icky."

"I know. But it's hard for me to believe a man—a man I thought I was in love with—could be not only a cheat but also possibly a murderer. I was the one who encouraged him to take the associate professor job at Colette. I must be the worst judge of character ever."

"You can't blame yourself for Alex's actions. He had us all fooled," Heather says. "Well, some of us weren't. Mustard never liked him. But think about it: What if Dr. Fontenoy figured out Alex used your research in the grant proposal and confronted him? Alex might have killed him to cover it up. Suddenly with the professor dead, Alex is a shoo-in for the department head. Now, that's what I call motive."

"No—" I shake my head vehemently. "It makes no sense. As for Mustard, Alex is allergic to cats. Although I suspect he made it up because he disliked the possibility of cat hair on his clothes. But the man hates confrontation. God knows he was shaking in his skivvies when I caught him with Phoebe. So, even if—and it's a big if—Professor Fontenoy confronted him, Alex would have lied his way out of it. So, as painful as it might be, Claire was right about him. Alex is much too smooth to be a scientist. That should have been a clue."

"Still, it must have something to do with your bees?" Heather says. "I mean—Fontenoy died next to your beehives, and Mother did say he wanted to speak with you."

"I'm sure it wasn't to beg me to resume my graduate program. We have to find out though—what was Fontenoy doing in the orchard?"

"Who knows? No offense, but entomologists tend to do weird things."

"None taken. I happen to agree."

Heather shrugs her shoulders. "Speaking of our mother, let me see your laptop. I've been dying to show you something I found online."

"Did you get any sleep at all?" Grudgingly, I slide my computer across the table. My reluctance is evident, and Heather raises an eyebrow in query. Of course, I hadn't closed my browser, so my cyberstalking of Sean Riordan remains on full display. "Don't say a word."

"Oh, I read the same article," Heather says. "He looks remarkably like his father, doesn't he? I wonder what happened in Afghanistan?"

"It doesn't say. Sean was a special investigator for the Army CID. Don't you watch NCIS?"

"Voilà! I discovered why Mother was eager to assist the new artist in residence."

The image on the screen is of an attractive man in a tight-fitted white T-shirt. He holds a mallet and chisel in his hands and poses next to a slab of gray stone. A thick mane of curly dark hair brushes his shoulders. The caption reads, "*Alberto Giovanni Vitale, contemporary sculptor, Milan, Italy.*"

"Oh, my!" I would put Alberto in his early forties, and his striking features and muscular physique rival an actual Renaissance work of art, let alone the sculptor. I whistle. "No wonder she was in such a hurry to get out of here."

"I knew she was up to something. Well, at least she's not bringing home another four-legged creature for us to care for." Heather sighs aloud.

"Meow." Mustard saunters into the room and heads for his empty food bowl.

"I didn't mean you, Mustard," Heather says. She wiggles her toes toward Lettuce, hiding under the table, who hops over to investigate. "Or you, Lettuce."

"Well, he has gorgeous hair. No one can ever accuse Claire of poor taste. Sometimes I'm not sure who the actual mother is around here. Her or us?"

"Oh, Indie," Heather says. "I know you're still angry with her. But maybe it's her way of coping with the divorce. As much as we hate it, it must be tough on Mom and Dad. They were married for more than twenty-five years."

"That's exactly my point. I always assumed our parents were happy. So, what happened?"

"They drifted apart," Heather says. "It makes sense when you think about it. Dad generated billable hours in private practice when we were younger, and Mom was off studying indigenous fiber traditions. They became two completely different people with no time for each other."

"Well, I blame Claire. She should have spent more time here at home instead of traipsing off around the world. I mean—who cares about that stuff?"

"That's not fair, Indie," Heather says. "You know, fiber art is Mom's passion. Perhaps she just wanted—or needed—to do something for herself. And besides, you weren't around to see how they were leading separate lives."

Deep down, I know Heather is correct. But the thought is still painful. In truth, I'd been too engrossed in my studies and, honestly, in my relationship with Alex. I'd supported his doctoral candidacy in more ways than I paid attention to my family. More's the pity. So, my parent's decision to end their marriage was a complete shock to me. One I have yet to reconcile myself with.

"Thankfully, we never had to choose between them," Heather says. "Anyhow, it's between them. Let's get your mind off it and focus on solving this murder."

"It hasn't officially been designated a murder. Maybe it was an accident?" Fat chance of that possibility.

"Indie, the man had a wad of honeycomb stuffed down his throat."

"I suppose it couldn't hurt to look into it." I'm reluctant to admit it, but Heather is right. I need to refocus my energy and prove my innocence. So, once again, I default to my comfort zone: science. "We need to form a hypothesis."

"Why?" Heather asks. "And how?"

To determine who, what, and where, we expostulate how a theory will turn out. Then we use scientific research, observation, and experimentation principles to determine a conclusion. Typically, we form a hypothesis. But in our case, the conclusion has been drawn.

"I'm almost afraid to ask. But the conclusion is—?" Heather asks.

"Professor Fontenoy's murder."

"I was afraid you were going to say that." Heather sighs. "So, how do we find the killer?"

"Essentially, we work backward from the time of the murder," I say.

"By—?"

"First, we determine a motive. Then we collect data on potential suspects. We'll need to delve into their backgrounds, confirm their alibis, and determine connections they might have had with Dr. Fontenoy. Hopefully, this will help narrow the suspect pool."

"Sounds a whole lot like snooping to me," Heather says. "And, you know, I'm very good at snooping. But what about the last thing you said—experimentation?"

"It's not snooping. It's research. Maybe we start by focusing on a few key individuals. By the looks of the injury to the back of Professor Fontenoy's head, I believe the probable motive of the crime was opportunity. Someone was angry enough to cosh poor Bob Fontenoy over the head with a brick. We need to figure out who. Then how can we elicit a confession." Yowza—what could go wrong with that? Just about everything.

"Whoa—wait a minute. That sounds dangerous," Heather says.

"The risk would be incalculably high. Therefore, I suggest we leave it up to the professionals."

"Perhaps we could come up with a list of suspects and observe them, you know, from a distance, but then turn our information over to the Sheriff's Department and let them take it from there." Heather shudders. "I don't want anything to do with getting a killer to confess."

"Neither do I." I can still picture the vicious gash on the back of Prof. Fontenoy's head.

"Oh, I know! We need a whiteboard," Heathers says, "and an easel and some of those yellow stickies. Maybe different yarn colors as they use on police shows."

"Are you talking about creating a murder board?"

"Yes, that's what we need."

"Great." That's not what we need, but I kept the thought to myself. "I think I'll ride out to Colette University this morning and talk to some of the people in the Science Department. You talk to the vendors here. Feel them out. See what gossip they may have heard. Or, maybe, you should get Granny Bea to consult the cards," I say in jest.

"That's a great idea," Heather says. "Almost all our artists are associated with Colette, and I'm sure Granny Bea will have some insight."

"I wasn't serious about the tarot consultation," I say. "And, by the way, I've been meaning to ask, what's up with her and all that pink?"

"Oh, she changes her color scheme according to her aura," Heather says. "I know you find predicting the future silly, but it's all about listening to your intuition. Bea is quite good at reading people. She was in such high demand at the Oak Grove Senior Apartments they were about to cite her for doing business without a license. So, we found her space, and Wyatt helped her set up an LLC. The Council on Aging also put The Hive on the senior citizen van's regular stops. You would be surprised how much money they spend here. Sabrina's Sensual is making serious bank off them."

It was difficult to clear the mental image of a bunch of blue-haired ladies ransacking the bins of Sabrina's intimate care products. "If you say so—you know I don't believe in all that woo-woo stuff. I prefer to follow the science and use logic to establish the facts."

"Intuition can be just as powerful as logic. Sometimes even more so," Heather says.

"Yeah, it does appear powerful emotions led to Professor Fontenoy getting murdered. Let's start with that premise." I fail to notice Heather staring at me with a pitying expression.

Chapter Six

I park in the lot closest to the Benjamin Franklin Mudge building, commonly called The Mudge. Appointed in 1864, Benjamin Mudge was the first Kansas state geologist. The more than a-century-old limestone structure makes up the entirety of the Science Department of Colette and houses the university's mineral and rock collections.

An amateur paleontologist, Prof. Mudge spent his leisure time excavating dinosaur fossils primarily from the *Permian* and *Mesozoic* eras. I bypass a collection from the species *Diplodocus,* skipping the opportunity to marvel at these Midwestern treasures and head straight up the wide marble staircase to the second floor.

As Chair of the Science Department, Prof. Fontenoy's office was at the back of a suite of rooms. I'm not sure whether that was by design or personal preference, but I suspect the latter. On a typical day, you pass a gauntlet of departmental assistants singularly determined to prevent you from speaking to the Chair. So, I'm surprised to find the offices deserted and make way unimpeded. Outside the open doorway of Professor Fontenoy's office, I hear the rustle of papers, a fair amount of sniffling, and what sounds like a muffled sob. I clear my throat in a poorly disguised attempt to alert the individual to my presence. "Hello," I call out and rap on the doorframe. "Anyone here?"

Janelle Grayson looks up at me from behind the large wooden desk piled high with books and papers. She's been crying—yesterday Bob Fontenoy was alive and breathing, and now he is dead. I know little to nothing about Prof. Fontenoy's personal life, but I assume he's left behind friends and family—people who mourn his loss. As the department's long-time administrative assistant, Janelle is one of them.

"You!" Janelle exclaims. She stands, and a large stack of papers immediately scatters onto the floor. "Now, look at what you've done! I just got these sorted." She bends down and begins to retrieve them. "How did you get in here anyway?"

The intensity in her tone startles me, and I raise my hands. "There was no one in the outer office. I knocked."

"I suppose they decided to take advantage of the university's offer of grief counseling and leave me here to sort through the mess," Janelle says. "And ah, uh—" her voice breaks. "I have planned poor Bob's memorial service." She sits down and begins to sort the papers into separate piles. "You shouldn't be here after what you've done."

"What—what have I done?" I ask, genuinely confused.

"What have you done?" Janelle repeats, "Ha! But, more likely, what haven't you done? You've managed to get the professor killed, that's what!

"No! Janelle, no. I didn't have anything to do with Prof. Fontenoy's death. You can't believe I would be involved in something like that—"

"I didn't want to," Janelle says. "I've known your family for years." She jerks her head back and stares up at me with wide eyes. "It's too hard for me to believe that the little girl I remember playing with her sister on the quad could be capable of something like this." She sniffs. "Or, maybe, as they said—you just lost your temper. Maybe—you—just snapped and killed him!"

Shocked by this accusation, I instinctively move toward her. "Who is saying that?"

"Everyone!" Janelle hisses. She picks up a hand stapler from the desktop and wields it like a cudgel. "Don't come any closer, Indigo Evans," she intensifies. "I will defend myself!"

"Janelle—" I hear the despair in my voice. This woman has been on the periphery of my entire life. It pains me to think she finds me capable of murder.

"And don't think I didn't give the police copies of those emails you sent to Bob," she adds. "They asked if I knew of any student angry enough to—to—" Janelle starts to sob, "—and yours came to mind."

That's one mystery solved. It was Janelle. The executive assistant always runs the department with zeal. "I'm not the only student who ever wrote an angry email to Professor Fontenoy."

"Well no," Janelle admits. "But yours was the only one who ever told him to go, you know, "blank" yourself."

The idea of the vulgar connotation from the mouth of a woman who teaches Sunday school at the First Methodist Church makes me wince. "In all fairness, I believe I used fairly benign terminology." I'd used the word "*schtup*," but today, I'm embarrassed by my crude implication. In my defense, I'd had cause.

"Same definition," Janelle says and sniffs. "I looked it up."

"It was wrong of me. As was my failure to factor in human emotions as a variable." I try to use a conciliatory tone. "But you must know how angry I was then."

"Janelle, I swear I had nothing to do with Professor Fontenoy's death. I'm here because Claire—um—my mother said Professor Fontenoy wanted to speak with me. I have no idea why, and I was wondering if you might know something. Anything?" My growing need to solidify Janelle as an ally is apparent. "Believe me. No one wants to find out who did this more than I do."

"Indigo! What are you doing here?"

Recognizing the imperious tone of Associate Professor Alex Carmichael, I fight hard to control my emotions. When I face him, I know him to be a liar, a cheat, and my former lover. The man is undeniably handsome in a Jude Law from *The Holiday* way. Like a romantic poet of old, a single curl of dark blonde hair drapes across his forehead. I've positive Alex secretly cultivates the image, engendering the fantasy of brushing the lock of hair back into place among his students, especially during those long, boring lectures on invertebrates. Didn't I say the man has an ego?

I may sound like a jealous ex. But I'm not. Time has a way of putting things in perspective. Presently, the impulse to yank that curl out by its roots is all I feel. Jealous? Hardly. Pissed off? Absolutely.

Alex must sense my intensity, and his blue eyes widen behind the lenses of black-framed glasses, yet another unnecessary pretension. Was there ever anything genuine about the man other than his vanity?

"Alex," I fight to keep my temper under control. "What are you doing here?"

"Alex, I found the tape measure." A new voice interrupts my observations. Phoebe Sutter's silhouette appears in the open doorway. Ugh. Can this day get any worse?

"I see you're measuring the place for new drapes," I say. "Isn't that a bit—premature?" Amused, I watch Alex take an involuntary step back and hear a nervous giggle from behind me. Good to know Janelle hasn't completely lost her sense of humor. Phoebe stands alongside Alex and slides one hand possessively into his.

Phoebe wears unrelieved black, and it's not in deference to the solemn occasion. Her inky black hair is the perfect foil for her round, dark eyes. Kohl eyeliner accentuates their epicanthal folds into a cat-like tilt. The black tank dress accentuates her slender figure, doing little to hide the elaborate serpent tattoo around her right bicep. An unpleasant memory stirs at the sight of the curving head and flickering tongue.

It's a memory of me pushing the door to Alex's office in the back of the entomology lab. I am curiously naive about the sound of soft moaning coming from within. Denial? Maybe. It is a classic defense mechanism. My hand reaches for the light switch, the harsh fluorescent lighting flooding the room and illuminating the naked backside of Phoebe Sutter. Well, it is bare unless you count the dental floss strip of her black thong as an item of clothing. She's crushed tightly against Alex's bare chest. The intertwined snakes inked along the base of her spine undulate in the age-old ritual of all rutting vertebrates.

I'm not proud of it, but the image still evokes a visceral response three months later. I fight hard not to let it distract me.

"Bookcases," Phoebe says. "Alex needs his books around him if he is to assume the extra duties of running the department with Professor Fontenoy gone."

I notice she didn't say dead and wonder if this is intentional.

"*Temporarily* assume, "Janelle says. "I see no need to bring in extra shelving for a tenure that will most certainly be temporary."

Phoebe slants a sideways glare at Janelle. "You don't know that."

"And neither do you." Janelle bristles. "You would do well to remember that fact."

"Ladies," Alex clears his throat. "Let's not fall prey to overwrought emotions during this challenging time."

"Oh, stuff it, Alex—overwrought emotions, my ass! You teach entomology, not Edwardian literature, and I'm sure you told the administration lies about me."

Phoebe starts to open her mouth, but Alex waves her off.

"The authorities insisted I speak with them. I had to tell them about your disagreement with the department. But I swear, I never mentioned you threatening me physically," Alex says.

"You mean when I threatened to cut off your gonads after I caught you and snake girl here doing the horizontal mambo on your office couch."

"For shame," Janelle hisses.

A slow flush suffused Alex's face. I'm surprised when he doesn't look away.

"That was most unfortunate. However, I have apologized to you and Phoebe for my lack of discretion. It was highly—unprofessional," Alex says.

"I can think of a few other unprofessional things you've done," I mutter. Despite Alex's denials, he hijacked my data into his project.

"It's alright, Alex. It's all water under the bridge. Some people need to learn when it's time to let go." Phoebe slides her free hand across his shirtfront possessively.

"Oh, believe me, I've let it go. My only goal is to prove you to be the fraud you truly are." I feel the blood begin to pound in my temples. "We both know where the bulk of the research came from on the Agro-Tech project."

"Indie, we've gone over this," Alex says, his expression pained. "This fixation of yours is becoming tiresome. It was a collaborative effort within the department. As the senior researcher in charge of the project, I was not obligated to credit you in a grant proposal submitted by the university."

"Credit me? I wasn't even given a footnote at the bottom of the page. I thought we were in a relationship until you took up with *Venom* here." I gesture towards Phoebe. "And besides, I read your proposal. You deliberately misrepresented my data. You skewed the results in favor of glyphosates. You, of all people, know I would never condone that."

"Ms. Evans, I wasn't aware you had re-enrolled at the university."

Dr. Victoria Medford, the current president of Colette University, enters the office followed closely behind by—yep, you guessed it—Interim Sheriff Sean Riordan. What had I said earlier about this day not getting any worse?

"I asked you a question, Ms. Evans," Dr. Medford repeats, her tone clipped.

"I'd like to hear the answer to that question as well," Sean adds.

For once, I'm at a loss for words, and my brain scrambles to come up with a logical explanation without sounding like a complete lunatic. "I—uh—"

"I asked her to stop by," Janelle says, stepping forward. "I've been reviewing Professor Fontenoy's papers and found a file containing Indigo's thesis premise. I thought she might like to have it back."

I have no idea how Janelle went from threatening me with a hand stapler to intervening on my behalf. But a bomb ticking down on a three-second timer could not have been more explosive than the effect of her words. Everyone stares at the folder Janelle holds in her hand with a mixture of avarice and fascination.

"I'll take that. I believe it belongs to the department," Alex says.

I notice he's managed to extricate himself from Phoebe's grasp. He steps forward, his hand extended.

Simultaneously, Victoria Medford says, "Ms. Grayson, surely you are not considering releasing university papers to a disgruntled ex-student?"

She gestures toward the file and makes as if to snatch it away from Janelle.

"Whoa, whoa, let's hold on here a second," Sean Riordan says. "As per your request, Dr. Medford, I've gotten a warrant to search the contents of Dr. Fontenoy's office. So, I'll take custody of this and all the other files here." He reaches for the folder. "Ms. Grayson."

Reluctantly, Janelle hands the file to Sean. We watch as he tucks it securely under one arm.

"I have two deputies on their way to transport everything to the Sheriff's Department's evidence room. After we finish up, I'll need one of you with administrative credentials to sign off on the inventory. Until then, perhaps you might be more comfortable waiting in the outer office while we process the warrant," Sean says.

The trio of Colette academics look askance at one another but dutifully turn to leave, led by Dr. Medford. As I move to fall in behind Janelle, Sean puts a restraining hand on my shoulder.

"I need a word with you, Ms. Evans," Sean growls.

His mouth close to my ear is not overly pleasant. On the contrary, it's his words that cause me to shiver.

"What do you think you are doing, Indigo?" Sean whispers.

Deputies Nolan and Hernandez walk into the room carrying two cardboard boxes and, without a word, begin retrieving files from the desktop.

"I'm doing what I must—finding out who killed Dr. Fontenoy 'cause it sure wasn't me." I pull away from him and stalk past the grim faces in the outer office.

"Indie, wait!" Sean says.

But I don't. I don't even slow down. By now, I've had it. I'm sick of the suspicions, the innuendos. I'm sick of the nightmare of seeing Prof. Fontenoy's face crawling with bees. But mostly, I'm sick of thinking one of the smug academics standing in the outer room could be responsible for his death. This notion propels me forward.

"Indigo, stop!" Sean commands.

I'm almost running down the deserted hallway leading to the stairs. As I grab the broad marble balustrade, Sean snags my shirttail and whirls me back toward him.

"Whoa," he says. "Slow down there. You could have fallen down the stairs. Are you trying to get yourself killed?"

I push my fists hard against the stiff front of his ballistic vest. "You said you believed me." I struggle to control my breathing. Now was not the time to panic. "Then why are you here, taking my files?"

"Shh—shush," he says. "I didn't just take yours. I took all the files. I figured we needed to lock down any potential evidence. The coroner has officially ruled Fontenoy's death a homicide."

"They did?" I feel him cover my hands with one of his own. It steadies me.

"And I do believe you," Sean continues. "Someone murdered Dr. Fontenoy in the middle of your beehives, and I'd be a fool if I didn't find the situation more than coincidental."

"What if it does have something to do with my research? Then I'm inadvertently to blame. So, let me help you," I plead.

"Oh no—no way! One murder in this county is enough. I don't want to have to worry about your safety as well. You need to let me do my job," Sean says.

"His death must have something to do with the research grant. Why else would Alex and Dr. Medford be so protective over those files?"

"You could be right. We'll have to see where the evidence takes us. In my experience, most motives for murder usually boil down to the base needs of love or money, and very often, a little of both," Sean says. "I'll tell you what, if I run across anything in those files I can't figure out, you'll be the first person I ask."

"I think we got it all, boss," Deputy Hernandez calls out from the hallway. "Dr. Medford signed the inventory sheet, and Dolan pulled the hard drive. The others want to know if they are free to leave."

Sean releases my hand and steps away from me. I feel the loss of his physical connection, and it shakes me more than I care to admit.

"Good job, Hernandez. Let me ensure Ms. Evans gets to her car safely, and I'll be right up," Sean says.

"It won't hurt them to sweat a few minutes," Sean mutters.

Chapter Seven

A fter my debacle in Detecting 101, I turn my new-to-me Subaru Outback south on Main Street, keeping the speed at a respectable twenty mph. My eyes are open for any passing Sheriff's Department vehicles. I may have escaped a murder charge, but I don't plan on pushing my luck. I know a few deputies who would love a chance to write me a speeding ticket. And when I say a few, I mean just one: Chief Deputy Harlan Pierce.

The late model Subaru was purchased from a graduate student last year who was headed to a research gig in the Galapagos Islands. He didn't see the point of paying for a vehicle abroad; I needed a reliable car. In retrospect, it was a wise decision on his part. Shortly after he'd departed, the Covid pandemic hit, and as far as I know, he is still hanging with the tortoises. The car combines rugged dependability and modern conveniences: all-wheel drive, automatic windows and door locks, heated leather seats, and plenty of cup holders. Best of all, unlike Gertie, my car has blessed, sweet air-conditioning. I crank it up to the max.

I can haul just about anything with the rear seats folded down. Well, anything except for a lit bee smoker. Currently, the cargo space holds my spare bee gear and six fifty-pound bags of llama and alpaca chow for Satin and Lace. Yes, those are their actual names, *à la* Roger Daltrey circa 1977. Claire was a fan. After dropping off the feed, I plan to swing by and check on a couple of packaged bees placed near the farm pond.

Evans Road becomes all gravel once you leave 187th Street. It's surrounded on either side by acres of corn and soybean fields. It dead-ends on a 360-acre plot once farmed by my grandfather and his father before him. Granddaddy Evans had been a kind and dedicated lover of the land. He never once complained about the harsh toll the Kansas elements took on his body and land.

At one time, my Grandma Nettie was as sharp as a tack. She had a good head for numbers and kept the farm afloat even in the lean years. But age is not always kind, and Grandma started to lose a step in her later years, developing dementia.

Once, she'd gotten lost driving to her job at the First City Bank of Colette. Unfortunately, she was ten years into her retirement from the job. A Kansas State trooper found her parked and disorientated at a toll booth some thirty miles outside the capitol city of Topeka.

Grandma Nettie would spend her remaining years in a memory care facility. After her death, my grandpa seemed to give up on farming and leased most of his tillable acreage to neighboring farms. Yet, he still maintained a sizable garden and a few beehives. As a young girl, my first taste of honeycomb was pure magic, and I'm sure my love for bees arose from those early memories.

From the road, my grandparent's original white farmhouse stands solid and welcoming in the grove of oaks my great-grandparents planted around the turn of the century. Miguel and Rosa Gutierrez and their two children currently occupy the space.

After Grandpa Evans passed, the farm was split between my dad and his older sister.

Aunt Rachael, a retired nurse, lives in Overland Park, an affluent suburb of Kansas City. OP in no way resembles the corn fields of rural Corley County. My cousin Jason, a second-year orthopedic resident, has no desire to become a small-town doctor. So, Aunt Rachael sold off her acreage, leaving my dad with the original farmhouse plus 110 acres.

While my grandpa was alive, Mom and Dad lived near the university. My mother still occupies the beautifully detailed craftsman bungalow we'd grown up in. As a Washburn Law School faculty member, Dad lives in a condo in Topeka, but he spends most of his weekends in a cabin built by the small fishing lake on the farm. Earlier this spring, I'd placed two new sets of packaged bees by the farm pond to forage on native flowering plants growing along the wood-line.

My dad and his student-led legal aid clinic represent Miguel and his wife, Rosa in their asylum flight from Venezuela. I am no expert on immigration law, but I know asylum can be difficult to prove. Miguel, a former economics professor from Caracas, had spoken out against the devaluation of his country's currency. Harsh government crackdowns on dissent and rampant inflation left the people of Venezuela unable to purchase essential goods and services. Those who were able to leave did so.

Through the aid of a local Catholic charity, the family was shuttled to Kansas. My dad's clinic took up their case. Currently, the family is ensconced as caretakers of my grandfather's farm while awaiting the court's decision. In the past year, they have lovingly repaired the old home to much of its former glory. The place has never looked better. I spy Miguel and pull up to the gravel parking area in front of the freshly painted barn.

"Buenos días, Miguel."

"Buenos días, Señorita Evans." Miguel says. "Còmo esta?"

"Muy bien." Reaching the limit of my Spanish vocabulary, I switch to English. "I'm sure the girls are annoyed with me." I nod at the two furry heads peering over the top of the metal pipe fencing. The act of popping the hatchback sets off a series of excited clucks from the taller one of the pair, a cream-colored, long-muzzled llama named Satin. She is followed by a low hum from the cherubic face of the smaller alpaca, Grace. Both *Camelids* and camel family members are highly prized for their fleece.

As you can guess by now, the pair belong to my mother. Claire adopted them from the Lucky Llama rescue during her Peruvian fiber phase. Until Miguel arrived, their daily care fell onto my sister and me. Well, mostly me.

"Sì, Señorita Indie, the ladies are indeed a bit—how you say in English? 'Peckish,'" Miguel says. "They are happy to see you."

"I'm sorry I didn't get by yesterday. But, unfortunately, there was an incident."

"Sì, I was sorry to hear about your troubles." Miguel says.

"Thanks, but finding a dead man beside my beehives is more than just trouble." The edge to my voice is unrelated to Miguel's innocent query. I know how bad news travels in a rural community.

"My apologies," Miguel replies. "Unfortunately, coming from my country, one can become callous to such things. But you are correct—a man's death should never be relegated to such a common description."

The clucks and hums of Satin and Lace intensify as he empties the chow into the large metal garbage can inside the barn door. Now it's my turn to feel chastised. I can't begin to imagine what he and his family have endured, leaving friends, family, and country. Yet, here I am feeling sorry for myself. You need to get a grip, Indie.

"I apologize, Miguel. I'm shaken up by the Professor's death being ruled a homicide. Murder is practically unheard of here in Corley County."

"Dios mío! Murder?" Miguel exclaims. "I did not realize. I thought we had left all behind us in Caracas. This place seems like a haven. But my children—should we be concerned?"

"No, no—" Despite my misgivings, I reassure him. "I'm sure it was an isolated incident. Something personal involving the Professor." And me, of course. I measure two healthy scoops of chow and walk over to fill the black rubber feeders mounted on the metal pipe railing. Taking a moment to enjoy the sounds of the pair's excited chittering, I rub the ears of Lace while staying clear of Satin and her propensity to spit at anyone who is near her food. "I appreciate you taking such good care of the girls."

"It's my pleasure. The ladies are no problem, except when they get low on chow." Miguel says, smiling. "I am sure they will be excellent guardians for your mother's new sheep. Especially Satin. She is quite protective, as you know."

"Sheep! What sheep?" Even as I ask, somewhere in the back of my mind, I recall a recent conversation between my mother and Mari, the student who mans Claire's hand-woven fiber booth. They'd been discussing sheep breeds and their differing wool properties. So, of course, I had tuned them out. "Please don't tell me my mother has purchased another lot of furry animals?"

"Sí, Señora Claire has purchased a dozen Tunis ewes for one of her upcoming fall classes," Miguel says with a smile. "I have been reading up on this breed. They are excellent wool producers and easy to raise. Llamas and alpacas make excellent guard animals for sheep, and I have promised Manuel and Maria a border collie pup to train. Manuel watches the herding dogs on YouTube. He is very excited. Who would have believed I would end up raising sheep on a Kansas farm one year ago? It's truly a miracle."

"If you say so," I murmur. "Miguel, I hope you know you aren't obligated to be involved in my mother's crazy schemes. Does my father know about this?"

"Oh, sí, he is fine with it. And in exchange for the care and feeding of the ewes, your mother will purchase the wool from us at fair market value. We can sell off any lambs next year for profit. The ewes are just a start. We will put the funds toward my children's college education."

"If you say so—better you than me, Miguel." Even after their divorce, my father still couldn't say no to my mother when she got a crazy idea in her head. "I'm going back to the cabin to check on those hives. They haven't been giving you any problems, have they?"

"No, both hives were active a few days ago. But the day before yesterday, I saw a vehicle over there, I did not recognize. When I checked the hives, I noticed one had little activity. Plenty of bees were around the second one, but it was close to dusk, so I assumed they might be inside. I knew you would come today and did not want to bother you. I hope they are okay?"

"Hmm—yeah, thanks. I'm sure it's fine, but I'll check on them. Did you see anybody else back there?" I'm more than a little curious.

"No," Miguel says, "but I could see tire tracks down by the pond. When I returned to the house, I saw the rear taillights of an SUV headed east on Evans Road. I think it might have been silver, like yours. At least, I think so. The gravel road was very dusty, and it was almost dark."

"A silver SUV? That's odd. I seem to recall I saw one somewhere recently. Maybe it belonged to one of the Muellers? They had purchased 250 acres from my Aunt Rachael. Bill Mueller, Sr. drives a white pickup, but I'm unsure about Junior. Still. "I've got my Bug Baffler in the car, so if need be, I'll peek inside the top lid of the hives."

"Perhaps, after yesterday, I should go with you," Miguel offers.

"No, I'll be fine. You have your hands full, getting ready for your new charges. It won't take long. I can always come back later if necessary need."

"Okay," Miguel says, a note of doubt creeping into his voice. "But you must be careful."

"I promise," I say. "I'll honk the horn three times to let you know all is well on my way out." I hop into my car and glance in the rearview. Miguel is watching, a pensive look on his face.

The cabin is actually what in this part of the country we call a "barndominium." These versatile, wood-framed, metal buildings range from a basic stall barn to tricked-out living quarters. Dad's cabin is 30'x60' with a two-bedroom, one-car garage layout and a bank of tall windows on the west end of the building, designed to view the glorious Kansas sunsets. The design is open-concept, and the interior is well-appointed. Ceramic floor tiles flow throughout the space, their neutral color broken up by various Claire's hand-woven rugs. Several of our mother's abstract wall hangings are mounted beneath the open wood beams, giving the place a cheerful rustic vibe.

Parking in front of the garage door, I traverse the cement apron beneath the overhang, spying a sheet of paper between the front door and the frame. That's odd. Miguel mentioned seeing a vehicle back here, but nothing appears amiss when I peer through the front window. I grab the paper and turn the knob. Locked. The glossy advertisement reads: "*Matthews Realty. Here to meet the needs of Corley County and Beyond.*" Beyond what? And more importantly, why is there a flyer from a real estate agency on the front of my father's cabin? The attached business card belongs to Brenda Matthews. On the back are handwritten words, "*Let's talk!*" underlined in purple ink. What the—?

Why would Brenda Matthews be driving back here on private property? Is Dad looking to sell the cabin or, more importantly, the farm? Even though I don't spend as much time out here as I used to, the thought of the old home place not being in the family elicits nostalgia. I have more than potential legal woes to discuss with my dad and make a mental note to call him ASAP. Still, I've wasted enough time focusing on things out of my control. I need to check on the hives.

Pulling on my Baffler and muck boots, I grab a pair of gloves and the hive tool I keep in the car for emergencies and skirt the pond to where I'd placed the two new hives of packaged bees. Packaged bees are a wood- and wire-framed box of starter bees with a fertilized queen. The average three-pound package contains about ten thousand bees.

The installation of a new package is fairly simple. You dump the box of worker bees across the top of a ten-frame Langstroth brood chamber. The deep wooden box sits atop a bottom board, two deep covered with an inner and outer cover. Initially, the queen is held separately in a tiny wood-framed box alongside a few attendants, plus a sugar plug. I had suspended the queen cage between two tightly fitted frames long enough that her pheromones were distributed throughout the hive, and the worker bees recognized her as queen. Then, ta-da, a new hive is formed. At last check, both hives flourished with a built-out comb on two deep brood chambers and their queens had established a good brood-laying pattern.

Approaching the hives from the back puts me in the best position to remove the top lid and inspect the frames without using my smoker. There is a lot of busy activity around the hive painted with honeysuckle, but the hive with the delicate trillium flowers is conspicuously quiet. I remove the two top brick weights and use my pry bar to loosen the lid. No bees crawl from the center of the inner cover. Next, I gently tap on the side of the box—still nothing. Finally, I use the edge of the pry bar on all four corners of the lid to break the propolis seal.

Typically, this action would bring about an angry horde of guard bees. This suspicious lack of activity is atypical. Once I remove the inner cover, I notice that capped honey is present around the edges of an empty pattern of hatched brood, but other than a few robber bees from the honeysuckle hive, nada.

Breaking the seal between the top and lower chambers, I remove the top box to inspect the second. Yet again, no bee activity. I examine both for damage or evidence of hive beetles or wax moths, two of the most common culprits. Again, everything appears normal except for the absence of bees, leaving me to surmise the bees have absconded.

What has made them leave? The question is one of those beekeeping mysteries without a definitive answer, but the result is the same whatever the reason: total abandonment of the hive.

I feel sick at the sight of the empty brood chambers. I should have been here yesterday. But, no, Miguel said it was the day before he'd first noticed the lack of activity. The same day he saw an SUV. I think again of the Matthews Realty flyer and the business card in my pocket from Brenda Matthews. Thanks to the interlude at the four-way stop sign, I know she drives a silver car. Could she have anything to do with my bees absconding?

No way. Brenda Matthews is a professional businesswoman. I can't imagine her going anywhere near a beehive. Maybe I'll have a chance to speak with Karl later, but I need to put this mystery on the back burner for now on a low simmer. I still have a murder to solve.

Chapter Eight

B esides the university, Agro-Tech Industries (ATI) is the county's largest employer. The plant runs two shifts from 0700–midnight and produces several popular herbicides/insecticides and a variety of soil conditioners.

Without getting too far in the weeds, pun intended, a soil conditioner's purpose is twofold: faster growth and better crop yield. An insecticide targets crop pests selectively. For example, chinch bugs, from the order *Hemiptera* are nasty little buggers that cluster at the base of the cornstalk and suck it dry of juices. They can quickly destroy a cornfield and are more prevalent in drought-prone years. Is this evidence of climate change? Maybe. But I wonder what the Kansas farmers who'd survived the 1930s Dust Bowl would have to say about today's weather?

Today, the most commonly used insecticides are non-selective and kill on contact. Bees included. Still, topical applications remain the most economical choice for most farmers. So, a product able to prep the soil and prevent the early onset of weed growth would give farmers a leg up during the relatively short growing season. Especially here, in the middle of farm country, volatile weather wreaks havoc on spring crops. And, due to a strong marketing campaign, ATI-EX50/50, marketed as Glo-Grow, a soil conditioner and herbicide combined, has become one of Agro-Tech's top sellers.

Now, I know you are wondering. Don't chemical crop applications equal lousy farming practices? And what happens to the bees and other native pollinators?

Why can't all farms be organic? Many small family farmers believe large-scale organic farming is not economically feasible and, over time, leads to diminished crop yields. Kansas Highway signs proclaim: "*Kansas farmers feed 155 people, and you!*" Agriculture, with forty-five million acres of tillable farmland, is the number one industry in the state. Kansas leads the world in exporting wheat and grain products. The demand has never been higher. So, for those concerned about pollinators' plight, and believe me, there are many, we are swimming upstream against a surging tide of billions of dollars in export sales.

Suffice it to say it crushed me to discontinue my fieldwork with honeybees. Pursuing a graduate degree was about something more than another rung on my academic career ladder. Agriculture and preservation can coexist. I'd viewed my fieldwork as a path to legitimate research opportunities. I'd wanted to make a difference. After all, no one is more concerned about the land's longevity than the farmer. Putting aside that I caught Alex shagging his teaching assistant, I'm still pissed off at him. He, of all people, knew I would never shill for the agrochemical industry. I used to think he wouldn't, either.

I approach the Agro-Tech entrance slowly. Several male employees appear to be consuming what's left of their brown-bag lunches. Others off to the side are smoking, hopefully out of proximity to volatile chemicals. What do you want to bet none of those brown bags contains a single veggie wrap or kale chip?

Suddenly, I'm starving, hungry enough to eat day-old vegan cookies if I had some. I doubt the portly security guard who has just stepped out of the guard shack has ever consumed one. Instead, he eyes my vehicle with suspicion. The guard's overt curiosity goads me to drive through the open gate. What reason could he have to be so curious? For all he knows, I'm an applicant seeking employment or delivering food for DoorDash. But, judging by the number of pickup trucks with Skoal stickers on their back windows and the fact we don't have food delivery out here in the county, neither is a possibility.

Engaging the driver's side power window, I pull up to the guardhouse. "Good mor—I mean —afternoon." Surreptitiously, I check the digital display on my dashboard. It flashes 12:00. Remember, Indie, bad things don't always happen at high noon.

"Afternoon," he says. "Can I help you?"

The guard, a man in his late fifties, peers into my window. His badly dyed combover is on full display. He wears a gray uniform shirt with prominent black epaulets and a gold name tag with "Henderson" pinned to his front pocket. The paunch of Henderson's belly squashes up against the door frame, causing the buttons to strain. His head tilts in this awkward angle a bit too long. I'm aware it's not my eyes he is looking at. Instead, I'm suddenly self-conscious over the transparency of my T-shirt, pulled tightly across my breasts, confined by the shoulder harness of my seat belt.

"Ah… yes… I want to see the manager. Mr. ahh— Mr.—?" I stall as I have no idea who runs the plant. "I can't seem to remember his name?"

"Carson Wells," Henderson says. "Wells is the plant manager, but he ain't here. But the shift foreman, Josh Blake, is in."

"I know, Josh!" I grasp onto the name like a lifeline. "We went to high school together. Is it possible to speak with him? Could you tell him Indigo Evans wants to see him?"

"Indigo? What kind of name is Indigo?" Henderson responds.

His words strike me as rude, and he leans further into my personal space when I don't reply.

"Yeah, I can get ahold of Josh for ya. I'll see if he has the time. You got an ID?"

Henderson's intrusiveness is getting on my nerves, so I pull out my defunct student ID from Colette with no vital statistics or personal information. I pass it through the opening and, as unobtrusively as possible, re-engage the power window forcing Henderson to step back.

"You've cut your hair," he says.

Yeah, and—? At least now, Henderson is upright. He squints down at the photo. It figures. He looks like the type of man too proud to admit he needs reading glasses. "Could you call him for me?" I plead to play nice even though it galls me. "I'm sure you are busy, and I don't want to take up too much of your time."

"Sure," he says. "Indigo—what kind of name is that?" He shakes his head before returning to the gatehouse for a hand radio.

From my parking place between two enormous Ram trucks, I see the lanky form of Joshua Blake striding toward me. My little car must feel intimidated, and I pat the steering wheel reassuringly.

"Indie," he calls out. "Girl, it's been a month of Sundays since I've seen you."

Such is my relief at the sight of his friendly face. I practically jump out of the car and throw my arms around him. A cacophony of catcalls and whistles erupts in the background. "Josh, it's so good to see you."

Embarrassed, Josh steps back from my over-eager embrace and stares down at me. "Whoa, girl, are you okay?"

I get an up-close, personal view of Joshua Blake for the first time in a long time. Gone is the skinny, mullet-headed kid who warmed the bench as a second-string high school football player. Instead, Josh's hair is stylishly cut, and he wears a pair of nicely fitted black jeans and a black polo shirt monogrammed with "Glo-Grow: The Future of Ag."

Josh's father, Buster, worked at Agro-Tech when it was still a part of the county's weed management program. Buster was a big, blustery man known for his propensity to overindulge in drink, which worsened precipitously after Josh's mom left them. I recall a pretty nasty incident from our senior year. In the final seconds of the homecoming game, Josh failed to convert his place kick into an extra point. We lost the game. Buster screamed so loud at his son he'd been escorted from the game by one of the off-duty deputies providing security. I'll never forget the look of utter humiliation on Josh's face. Life within the Blake household could not have been easy. Yet today, Josh's blue eyes sparkle with good humor, and his smile is genuine. I find myself smiling back, happy to see him.

"Yeah, I'm fine. It's just been a rough couple of days, and the guard at the front gate, Henderson, gives me the willies." Then, out of the blue, I feel tears well in my eyes. I'm usually not a crier, but the flood of emotions I've been holding back in reaction to Prof. Fontenoy's death threatens to overwhelm me in the face of Josh's warm greeting.

"Who? Bud?" Josh asks. "Don't pay him no mind. He's harmless. A bit of a jerk, but otherwise harmless." He must have noticed my tears and concern flits across his face. "I'll have a word with him, Indie. And if you want, we can file a complaint with HR," he says. "I can take you there right now."

"No, no, Josh, there's no need for that. I'm just—having a bad day—but it's nothing I can't handle. I'm just happy to see a friendly face." Embarrassed, I dab at my eyes. "Look at you——shift foreman, huh? That's great. By the way, I was so sorry to hear about your dad's passing last year."

"Hey, eventually the booze catches up on your liver, but thanks. So, yeah, can you believe it?" Josh says. "I guess our chem teacher, Mr. Jackson, was wrong about me. Mixing random stuff in the lab finally paid off, and now, they pay me for it. But, hey—what's this I heard about a professor from the college killed by your bees? And they say you found him? What in the world was he doing out there? Is it true his face was eaten off?"

Just like that, Josh reverts to the persona of the seventeen-year-old boy I remember fondly. "First, Mr. Jackson was worried you might blow us all up. Second, the victim's face was not eaten off by bees." All true. A human apex predator most certainly killed Prof. Fontenoy, but I keep the knowledge to myself. "You've been misinformed, Josh. Where did you hear that kind of gossip?"

"Oh, I don't know—around." Josh waves his hands in a circular motion. "You know how everybody knows everyone else's business around here."

"Yeah, I know." It was one of the downsides of living in a small community. The upside was old friends like Josh. People with a sense of shared history and a willingness to talk. "Hey Josh, do you have a few minutes to speak with me about your new product, Glo-Grow?" I point at the monogram on his shirt.

"Sure," Josh says. "What do you want to know? Not much to say other than we've been getting a ton of new orders. We ship the stuff out day and night, even running two shifts, and we can barely keep up with demand. It's become a real boon for the company." He lowers his voice and adds proudly, "Don't tell anyone, but they even let me mix it sometimes."

Harking back to our chem labs days, I think this probably isn't a wise decision. "Tell me more."

"I can, but how 'bout I show you instead," Josh says.

Even better.

The Agro-Tech facility is a big, open warehouse with lines of massive stainless-steel tanks throughout. Overhead, a maze of metal piping runs above the containers. Flexible arms swing from tank to tank in timed intervals with brass spigots filling the barrels. Forklifts beep as they speed about, swapping full tanks with empty ones. The constant clang of metal on metal echoes in the open space, even against the roar of the enormous ventilation fans above us.

It appears to be utter chaos to an outsider, but the employees exhibit an air of purpose, and the facility is immaculately clean. Nothing seems remotely nefarious about the place except for the smell, which is noxious and overwhelming. All this busy activity must equal umpteen dollars for the company. *Ka-ching*. Good for you, Josh.

Josh is both effusive and informative. He answers all my questions about formulation and distribution. However, Josh clams up each time I bring up the absent plant manager, Carson Wells. No, he doesn't know where his boss is today. And no, he's not familiar with the terms of the grant Colette University received from his company.

"That's way above my pay grade, Indie," Josh says.

Josh insisted I wear the requisite *Tyvek* coverall, hair net, and paper booties. We'd picked our way among the vats of hazardous chemicals. The employees were predominantly male, and I garnered a few curious stares.

As a matter of record, I only met another woman once I got to the administrative offices. I'd commented on this, and Josh replied, "We've tried hiring a few ladies for the floor, but they always transfer to the office. They can't take the smell. There's not enough Axe body wash to get the stink out of their pores, so I can't say I blame 'em."

I can't say I do either. Once back outside, I also need to shower, so I power down all the Subaru's windows to allow fresh air. Bud sticks his head out of the guardhouse as my car passes back through the gate. I'm tempted to give him the one-fingered city wave but decide to behave and turn to drive toward town. Unfortunately, my plan to stop at Karl's will have to be postponed until after I've showered.

Back at The Hive, I find Heather engaged in an intense conversation with Granny Bea. Wyatt stands awkwardly to one side, arms folded across his chest. He watches the back and forth with the devotion of a lapdog. However, the diminutive figure of Beatrice Kowalski dressed in an eye-piercing yellow caftan commands my attention. I guess today's aura must be one of a banana.

"An exchange between a reader and client is strictly confidential," Bea insists.

"But Granny Bea, this could be our first real clue. You have to tell us what she said," Heather pleads. "Or at least give us a hint."

I wade into the fray. "What, who said?"

"Phoebe Sutter." Heather looks up, her eyes alight with excitement. "Phoebe came by the shop this morning. Cool as a cucumber, she waltzed in here and then headed straight back to Granny Bea's."

"What did she want?" Upon closer inspection, I notice our resident fortune teller's caftan is printed with large, yellow—are those pineapples? Perhaps she is channeling her inner Carmen Miranda. For those who don't know "The Brazilian Bombshell," look her up. Those fruit hats she wore were spectacular.

"As I was saying to your sister," Granny Bea replies, "it would be completely unprofessional of me to reveal the confidences of a client reading."

Before I can muster a rebuttal, Bea holds her palm up, facing outward. The gesture and the bright yellow outfit remind me of a school crossing guard.

"However," Bea continues, "I don't see why I can't give you my general impression of the young woman."

"Please do!" Heather urges.

"Oh, yeah—yes, please do," I agree with little enthusiasm. Seriously. Has it come down to this? Waiting to hear the impression from a woman dressed like Charo. "But how can you ascertain Phoebe's aura? She never wears anything other than black."

"Oh no, my dear, dressing in all-black is simply a way to disguise the color of one's soul," Bea explains.

Oh, brother. Seriously?

"What color is her soul?" Heather asks eagerly.

"Ebony, cinder, brimstone?" I retort.

"Indie," Heather chastises. "Granny Bea will never help us if you keep this up."

"It's alright, dear. Your sister's protestations don't bother me in the least," Bea says.

"Alright—please, Bea, tell us your general impressions."

Bea nods her head in a regal manner. "I feel the young woman is conflicted, battling with spiritual forces who seek to suppress her inner light. She desperately tries to conceal her true nature."

We all stand there momentarily, allowing Bea's revelations to sink in. Heather's mouth forms a round O. Wyatt shifts from side to side, looking supremely uncomfortable.

"That's it?" I ask.

Bea nods again. "Yes. I felt nothing sinister emanating from Phoebe Sutter. Although, I will admit those horrible snake tattoos are quite off-putting."

For once, Bea and I agree on something.

"So, you're saying you don't think she's capable of murder?" Heather asks.

Bea waves her arms in the air. "Isn't everyone given the proper motivation?"

What a load of claptrap. For Phoebe's unusual cosmetic choices, and believe me as a scientist, they are wrong in so many ways, she is a brilliant researcher. There must be an ulterior motive for her visit to The Hive, and I'm pretty sure it did not involve having her cards read. No, Phoebe Sutter is up to something.

"Nice outfit, Granny Bea. Wait—don't tell me—someone we know is going on a tropical vacation?"

"One never knows, my dear, but I did draw the Page of Cups this morning. So, new and unexpected opportunities abound," Bea says.

Oh great, here we go again. I refuse to take the bait, so I turn to Heather instead. "I've been trying to call Dad. Do you have any idea where he is?"

"Hawaii," Wyatts interjects. He slides his phone from his back pocket. "Correction, according to Flight Tracker, he's still en route. Which is probably why you haven't heard from him."

"Indie! Don't you remember Dad's golfing trip? That's why he designated Wyatt to represent you," Heather says. "Wyatt plans to touch base with him just as soon as he lands."

"Dad's trip was this week?" Disgusted, I slap my palm to my forehead. "I forgot all about it." Our father was meeting a few of his law school alums for a week of golf on the Big Island. It completely slipped my mind. Geez, I can't imagine why.

"By my calculations, he should land in just under two hours. I'll give him an hour to get to his hotel, and then I'll call and update him on your case. Is that okay, Indie?" Wyatt asks.

"Yeah, I guess so." I can't believe I forgot about my father's trip. I need to speak with him, and not just about my case, thinking about the Matthews Reality sign. The chemicals at Agro-Tech must have affected my powers of recall. "I'm gonna go take a shower and call him later."

"I wondered what that smell was." Heather wrinkles her nose and fans the air in front of her face. "But don't wait too long to call," she adds. "You know how Dad is when he gets around his law school buddies. Before you know it, he'll be sitting at the bar with a round of tropical fruit drinks, the ones with the little umbrellas and slices of fresh pine—pineapple—" Heather's voice trails off, and her mouth drops open.

Heather, Wyatt, and I look toward Bea, resplendent in her pineapple-print dress.

"The Suit of Cups has to do with love and emotions," Bea pipes up. "Maybe your father will discover a new love on his tropical vacation? Wouldn't that be wonderful?"

She looks at my sister and me expectantly. When neither of us replies, Bea shrugs and shoulders a big bag from the counter.

"Well, I'm late for a smudging down at the Oak Grove Apartments." Bea lowers her voice to a conspiratorial whisper. "Mr. Hardiwick passed yesterday, and some tenants are afraid he's left behind some negative energy."

I sigh and stretch my neck from side to side till I hear a pop. "I do need that shower."

Chapter Nine

"Feel better?" Heather asks, entering the loft space. She sniffs the air. "You certainly smell a lot better. What was that odor?"

"I feel much better. It's the chemicals used to formulate the products Agro-Tech sells. I took a spur-of-the-moment tour of the facility this afternoon. Do you remember Josh Blake from high school? He's the shift foreman, and he showed me around. He is still a sweet guy. But, shift foreman? Yikes. Don't you remember when we had to evacuate the school after Josh almost blew up the chem lab?" Sunlight floods through the line of floor-to-ceiling casement windows fronting the main street of our apartment. I glance down at my watch. "Is it time to close already?"

"Not quite," Heather says. "But I closed early today. There is a Chamber of Commerce meeting at City Hall this evening. Angela and I volunteered to provide refreshments and must be there beforehand to set up the coffee and cookie tables."

"Vegan cookies, again?"

"We always offer several options, vegan being one of them," Heather says. "Why?"

"Well, if Sean Riordann shows up, ensure he gets the vegan ones. I think he really likes them." *I smirk.*

"Haha—very funny. You should come. It's going to get pretty interesting," Heather says. "Especially since you've toured Agro-Tech. Management is pitching a plan to expand the local facility. It's got both pro and con people stirred up. I hear the bigwigs from the Midwest regional office are flying in from Cedar Rapids."

"Management? Exactly who might that be?" My ears perk up at the possibility of meeting the elusive executive.

"Carson Wells is presenting in favor of the company," Heather says. "But Josh Blake is always somewhere around Carson. I believe quite a few employees will attend tonight's meeting. The proposal could add over two hundred jobs to the area, so it's a huge deal for Colette."

"How come I'm just hearing about this?" I ask.

"It was originally scheduled for last quarter, but the mayor and the council have tried to keep it on the down low. They tabled the discussion until the end of the spring semester to avoid any possibility of student-led protests. But a little bird told me a climate group from Topeka got wind of the meeting and plans to show up tonight." Heather's voice fills with excitement. "I'm looking forward to the fireworks."

"What little bird?" I raise my eyebrows in query. "Heather Grace Evans, are you stirring up trouble in our peaceful little hamlet?"

"Who me?" Heather purses her lips and whistles a trill-like bird call. "At issue is the plan to produce more toxic chemicals in our area. Someone should speak in opposition. As a de facto member of the Chamber, I've been encouraged not to voice my opinion at a public forum. I'm supposed to let the public decide."

"Meow." Mustard comes padding out of the kitchen. He looks up at Heather and winds his body between her feet. "Meow."

"Sorry, Mustard, there's no bird in here," Heather says. She picks up the cat and hugs him against her chest. "My sweet boy—did mean ole Indie not feed you?"

Mustard yowls louder. The cat levels his golden eyes on me, and I swear I hear the words, "Feed me."

"Okay, I will, and perhaps, I will come to the meeting. It could be a good place to consolidate my suspect list."

"You have a list?" Heather asks, her eyes alight. Mustard leaps from her arms and races toward the kitchen.

"I'm working on it, and the meeting is as good a place as any to consolidate it."

"You think the killer would have the nerve to attend the Chamber meeting?" Heather shudders. "Ugh, that's ballsy."

"I wouldn't be surprised. But sometimes it's seeing who is not there."

Residing in a loft space atop an old building constructed around the turn of the century has many upsides. The lofty ceilings with ornate crown molding accommodate period glass windows, letting in a ton of natural light. The feeling of history and solid craftsmanship surrounds you. However, dealing with antiquated electrical wiring and plumbing is a constant battle. As a result, Heather and I never do the dishes and take a shower at the same time.

Heather takes her turn in the shower, and I hear the pipes rumble ominously as the old boiler in the basement tries to heat the water and force it up three floors. The reverberation is an indicator the hot water is running low. Oopsie, my bad.

I pick up Mustard's food bowl and refill it, surprised when he doesn't immediately start to gobble down the tiny amount of weight-proportionate cat chow. Rather, he fixates on the space beneath Lettuce's hutch. The rabbit hutch is elevated about four inches off the floor, leaving enough space for a small rodent to hide. Great. I wonder if Mustard has cornered another mouse. Thus far, none of my humane traps and determents have kept mice from our space. No doubt, the abundance of free rent and food attracts them.

"Slacking on the job again, Mustard?" The cat does not move, and a low growl emanates deep within his chest.

"What's under there?" I'm not expecting an answer, but, to my surprise, Mustard turns his wide, unblinking gaze toward me, and I notice the long yellow hair standing at rigid attention along his spine.

"Grrr... grrr..."

The hutch is a two-level rabbit condo custom-built for our space. The bottom level holds a water bottle attached to the outside wireframe. A feed bowl filled with rabbit chow and chopped veggies sits inside. I recognize this morning's breakfast from the rough chop of the carrots. Hey, I'm no sous-chef, but the food remains untouched. "Lettuce," I call out, now slightly alarmed. Squatting, I peer into the burrow on the second floor of the cabinet and see Lettuce's long whiskers sticking out from the opening. His soft little nose wiggles frantically.

"Lettuce? What's wrong, buddy? Are you trying to tell me something?"

"Grrr—hiss—" Mustard lashes out, claws extended, and swipes beneath the hutch.

Now, I'm down on my hands and knees, peering into the dark space. My left cheekbone is pressed against the floor, and I sweep my cell phone light from one side to the other. All I see are fluffy rabbit and cat hair tufts, a not-unexpected consequence of living with pets. Then, in the far-right corner, I notice a dark lump of something. I refocus the beam, and twin pinpoints of red reflect back at me. What I assumed to be a mouse suddenly extends lengthways and slithers to the opposite corner.

"Yikes," I shout, jerking my head back and bumping into the water bottle. It falls onto the floor and drenches my lap. Mustard scrambles backward. I hear his claws gain purchase on the wood floor as he leaps for the kitchen table.

"Yowl—" Mustard cries.

Crap on a cracker. A mouse would have made a break for it before being picked off by the agile Mustard. Instead, the cat growls again, and the hair on the back of my neck stands upright at the low, feral sound. Mustard swishes his tail angrily from atop the relative safety of his perch. He doesn't seem inclined to abandon it. Meanwhile, poor Lettuce has burrowed back into the deep recess of his den.

"What's wrong with Mustard?" Heather appears in her bathrobe with a towel wrapped around her hair. She starts to move toward the irate cat.

"Yowl—"

"Heather, give me your towel!" I try to convey a sense of urgency in my tone.

"What—why?" Heather asks.

She starts to speak, but the expression on my face and the persistent howling from Mustard causes her to think better of it. Instead, she unwraps the towel and tosses it toward me.

"Don't make any sudden moves," I whisper, pointing beneath the bottom of the cabinet. "Just find me a sturdy cardboard box with a tight-fitting lid.

Heather looks at the space beneath the hutch, her eyes large with comprehension. She starts to retreat but hesitates at the doorway. "What about Mustard and Lettuce? Are they going to be okay?"

"Geez—thanks." I notice there is no concern for me whatsoever. "They're fine. Neither one of them is going anywhere." Or being of any help to me. I roll the damp towel up lengthwise. To seal off the escape hatch for whatever lies underneath, I stuff the towel between the floor and the bottom of the cage. I don't think Lettuce plans to appear, but to be safe, I latch the hutch door.

"Go get the box, and make sure it's deep enough. And while you're at it, bring me a broom."

Heather returns with a box and a broom. I notice she's pulled on a pair of tall, rubber rain boots. Now would be a good time to make a snide comment about her appearance, but I think better. The box, the broom, the bathrobe, all combined with her wild wet hair—it would be too easy.

"Don't freak out, but I'm pretty sure it's a snake under there."

"Oh God—a snake?" Heather puts one hand over her mouth and moans, "You know, I hate snakes."

"I know you do." Despite obtaining an undergraduate degree in entomology, I don't care much for them either. But now is the time to remain calm. "I need you to help me slide out the hutch while I corral it into the box."

"Okay," Heather whimpers. "What do I have to do?"

"I'll use the towel to create a funnel between the snake and the wall. Then, hopefully, we can force it into the box with the broom."

"That's your plan? Maybe we should call in an expert?" Heather says.

"I am the expert. We can do this." Broom in hand, I use the handle to push the rolled-up towel further under the cage, making my movements slow and unthreatening. I hear the creature hiss and bump up against the terry cloth barrier.

No. On second thought, the hissing sound is coming from behind me. Mustard has his round rump planted firmly on the table. He cleans his paws and stops to hiss at me as if this were all my fault. Yep, I should replace him with a dog. Heather remains frozen in place; the curls on her head start to dry and frizz. The incongruity of the scene is so ridiculous I could laugh if I weren't such a scaredy cat. *Hello, my name is Indie, and I am an entomologist afraid of snakes.*

"What should I do with the box?" Heather asks.

"Place the widest side flat on the floor with the open end facing the hutch. Then unlock the front castor on the hutch. Once everything is in position, I'll push the opposite end of the towel toward the wall, and we'll pull the end of the cage out to create the funnel. I'll use the broom to sweep the snake into the box."

"What if it jumps over the box?" Heather asks, a slight note of fear in her voice.

I hadn't thought of that scenario. Some snakes, like the *Chrysopelea paradisi,* are known for their ability to leap long distances. Native to Indonesia, they can jump or glide from tree to tree. However, I am relatively confident an exotic paradise tree snake isn't hidden beneath a rabbit hutch in Colette, Kansas. At least, I hope not.

"It's not going to jump. Remember, snakes are more afraid of us than we are of them." So says every *ophidiophobic* person ever.

"Easy for you to say," Heather moans.

"Pull one of those kitchen chairs to stand on, then slide another over for me. I'll leverage the cage with my broomstick from the top of the chair." And, fingers crossed, avoid the possibility of a snake slithering across my bare toes.

My broom at the ready, I watch Heather position the box and disengage the caster's lock. "You're doing great—just slide out the front corner of the hutch. Good. Now, go jump on top of the chair."

I wait until Heather is safely in place. I notice her chair wobble. Mustard pauses mid-cleaning at the sight of his person balanced atop a perch closer to the action.

"Meow," Mustard cries. He squats back on his haunches as if preparing to spring.

"No! Mustard—don't!" Heather shouts.

They say two things can be true at once. For Mustard and I, the saying is apropos when we both try to jump simultaneously. The word "try" is emphasized because neither of us succeeds. Mustard catches the edge of Heather's chair with his front claws. Although valiant in his effort, Mustard's excess flab and the force of gravity are his downfall. The cat lands on the floor with a thud, causing Heather's chair to slide sideways into mine. Her chair teeters precariously as she leans forward to rescue her feline friend.

For my part, I make it to the top of my perch and manage to poke the broomstick into the crevice behind the rabbit hutch. However, I overestimated the leverage needed on my makeshift pry bar. The corner of the cage slams my chair into Heather's. The snake uncoils, and I catch a glimpse of red, black, and white stripes before it starts to scale the wall. Heather and I both crash onto the floor.

"Snake! Climbing the wall!" I scramble, quickly right the chair, and clamber back onto the wooden seat. Then, out of the corner of my eye, I see a flash of gold as Mustard streaks from the kitchen. The cat is followed closely behind by Heather, boots clopping and wild curls streaming down her back.

Silence. Not a sound comes from inside the hutch or beneath it. I take a deep breath, trying to understand what just happened.

"Red touches black, safe for Jack. Red touches yellow kills a fellow."
Phew—I'm pretty positive I saw red touching black.

"Indie," Heather calls out. "Are you okay?"

Heather's voice is muffled and well beyond the kitchen doorway. "Yeah, I'm fine. I'm sure it's a Central Plains Milk Snake. It's quite common and nonvenomous. And judging by what I glimpsed of its size, relatively young. Where are you?"

"On the back of the couch. I have Mustard with me, so he's safe," Heather says. "And when you say 'sure,' exactly how confident are you?"

Gee, thanks. It's such a relief knowing Mustard is safe. "Sure enough, I got an A in my snake identification class." Of course, that's a lie; there was no snake identification class, but there should have been. "Just stay where you are. Both of you."

"Okay," Heather agrees.

My sister's response has no hesitation, so I'm on my own. Peering over the back edge of the hutch, I expect to see a coiled-up snake pressed into the corner. To my surprise, the space is empty. Uh-oh, milk snakes are known to climb but scale a twelve-foot ceiling? Hardly. I look up anyway. No snake. Okay, Indie, now what?

"Indie," Heather calls out. "Is everything alright? It's quiet in there."

I look down into the narrow triangular space leading into the box. Could it have been this easy? I swipe the broom bristles before the box's opening and feel something strike the edge. Eureka, I've already got it. Now, what am I going to do with it?

"Heather?"

"Yes," Heather replies.

"Please call the Sheriff's Department."

"It's about time," Heather says.

My biggest complaint about TV cop dramas is that the same two stars respond to every call and solve every crime like no other policemen are available in the city. Well, for us rural folks, it's our daily reality. Sean Riordan is ushered in by Heather, who I can't help but notice has changed into jeans and a cute crop top. I see Harlan Pierce swagger in, following closely behind him. If you recall, I'd just gotten out of the shower at the beginning of this little escapade. No, I am not standing atop a kitchen chair naked. God forbid. But I am wearing an oversized, worn, green jersey emblazoned with the words "Colette Saints" and a pair of cotton granny panties. What—they're comfortable, and at least they are clean. Of course, I'm barefoot and need a pedicure, but at least I'd had the foresight to shave my legs.

"We do need to stop meeting like this, Indigo Bleu Evans," Sean says, a big grin plastered on his face.

Harlan chortles. "Hold still, DB. I need photo evidence of this. The guys at the department will never believe this one."

Chapter Ten

"**B**eekeeping and snake wrangling. What other talents are you hiding from me, Indigo Evans? Please say it's gourmet cooking?" Sean asks.

"Ah, cooking would be a big NO. I can barely boil water—ask Heather. She's the chef in the family." I'd followed Sean downstairs after we secured Stripes in the box. I'd taken to calling the milk snake by name in a show of bravado, not wanting to divulge my ophidian phobia.

As was his custom, Harlan wanted to, in his words, "exterminate the sucker." But Heather and I protested so vigorously that Sean had to intervene again. Finally, the compromise of catch and release was agreed upon. Sean and I would drive Stripes out of the city limits. Meanwhile, Harlan was dispatched to city hall, ostensibly for crowd control but really to help Heather and Angela set up the refreshment tables.

"I've never ridden in a Sheriff's Department vehicle before," I slide into the backseat and hold tightly onto Stripes' box.

"You can sit up front, you know. You're not a suspect," Sean says.

"I think I'll stay back here with Stripes," I say, keeping my hand firmly on the box. "At least until I know he is safely released. We'd never catch him if he gets loose in here."

"Good point," Sean says.

Unlike Heather, I am not the chatty type. I tend to shut down when nervous, and we settle into silence as the SUV pulls slowly onto Main Street. Sean adjusts the rearview mirror, and I see his vibrant blue eye reflected on the right side of his face; the black patch again obscures the left.

I may not be much of a talker, but I blurt out my thoughts at the most inappropriate of times. "Does your eye bother you much?" Feeling clumsy, I try to rephrase my query. "I am so sorry. You don't have to answer that. It was rude of me to ask. And, if I didn't say it before, thank you for your service." Stupid, stupid, stupid.

From behind, I watch Sean roll his shoulders. Finally, he tilts his head to the side and looks at me in the mirror with a grin.

"You should see your face right now," Sean says. "It's all red and blotchy with embarrassment."

I can only imagine how I must appear to him in my current state. At least I'd exchanged my running shorts for a pair of jeans and grabbed the muck boots from the back of the Subaru. After all, one cannot wrangle snakes into the wild barefoot.

"And to answer your question, no longer," he says. "It was my honor to serve."

"You were with the Army CID?"

"Yeah, I was. How did you know?"

"Google." Somewhat reluctantly, I admit to my cyberstalking.

"Oh—" Sean says.

Quietness once again envelops the space as Sean cranes his neck to make a left-hand turn off Main Street, heading out of town. He accelerates the SUV and glances back at me in the mirror.

"Before the withdrawal, I was on assignment investigating a green-on-blue attack in Afghanistan. On the way back to Kabul, our vehicle hit an IED. My driver was killed, and my partner broke her back. A piece of shrapnel punctured my eye and ruptured my eardrums. My hearing came back, but my vision didn't," Sean says. "But I consider myself one of the lucky ones."

"Oh my—I am so sorry. I should have never brought it up."

"It's okay. It's natural to be curious. I'm alive, and I've learned to live with it." He points to his eye patch.

"You said your partner broke her back? Is she okay?"

"Well, if you call crushing your L5 and 6 in your spine 'okay.' But Lainey has regained much of her primary function and taken a job with the Tucson P.D. She's still on desk duty but is one hell of an investigator and an even more determined patient. So, I'm confident she'll return to the field soon," Sean says, admiration in his voice.

"Thank goodness. This Lainey sounds like an exceptional woman."

"Oh, she is—and wife to a fellow cop in Tucson and a terrific mother to two little girls. She's like a big sister to me." He glances up again. "In case you were wondering?"

"Oh—" I feel myself blush. I was wondering but I'm too embarrassed to admit so. "You said your driver was killed? That's awful. War is so senseless."

"Yes… it can be." Sean hesitates. "Private First Class Jonathan Dooley, from Boseman, Montana. By all accounts, a good kid from a nice family."

"You didn't know him?"

"No. I was never fortunate enough to meet PFC Dooley until the day he was tasked to drive Sarah and me back to the airport."

He must have noticed the stricken look on my face.

"I know to you it might sound cavalier. But you see, Indie, I saw the absolute randomness of these incidents during my years in the military. It's never fair or equitable. And if I've learned anything in my thirty-two years on this planet, we shouldn't take life for granted. So, I try and live my life mindful of the sacrifices of good men and women like PFC Dooley."

"How? By being a policeman here in Colette, Kansas?"

"By taking one day at a time," Sean says mutedly. "By remembering to appreciate the beauty in the things around me. Like—"

"Like what?" I urge.

"Like the sight of the sun setting over the Flint Hills. Or taking a moment to appreciate a full moon on a clear night, a million stars above you—with the knowledge there are good people, good soldiers standing guard all around the world looking up at those same stars. And those same stars will be there when we're all gone. That's when you realize—the universe is bigger than you, and suddenly, your problems seem insignificant."

Once again, the familiar wry smile crosses his face, and I feel my pulse flutter.

"Even if it's in Colette, Kansas. Even when rescuing a beautiful, complicated lady from a snake under a rabbit cage," Sean says.

"I wouldn't exactly call that a very beautiful experience. Harlan about laughed his ass off," I say.

"Why do you always turn a compliment into a joke?" Sean asks.

"I guess I use sarcasm and humor to deflect." And hide my inadequacies. I don't say the quiet part out loud.

"Well, I must say, finding you on a chair with a broom was pretty humorous. But, hey, you've been unusually quiet about Stripes there. Don't you have any theories on how a snake made it up three flights of stairs to hide beneath a rabbit hutch?

"I might have a couple." I have exactly one and it has to do with the timing of Phoebe Sutter's visit to our resident palm reader. But, so entranced by Sean's revelations, I'm reluctant to break the companionable truce between us. Suddenly, the radio squawks, and we turn onto a familiar gravel road.

"601, what is your 10-20?"
Sean cues his shoulder mic. "Evans Road with 10-11."

My head pops up as I look out the tinted window. "Hey, we're on Evans Road, heading toward our farm."

"Yep, I figured the fishing pond on the backside of your dad's property is as good a place to release your snake friend."

"How did you know about the pond?" I ask.

"You're not the only one who knows how to use Google," Sean responds. He reaches for the mic again. "10-23, this is 601, Code 7."

"What do all those numbers mean?" I ask, puzzled.

"It means I'm taking a break with a woman and a truant snake. Isn't that how the world was introduced to original sin? If I recall, it didn't work out well the first time, so please tell me your dad hasn't planted a garden of apple trees back there?"

Oddly charmed by his biblical allusion to Eve eating the apple, I laugh out loud. We turn onto the drive, and I glimpse my grandparent's house in the waning light of the summer sun. The image of an enduring home place evokes potent memories of family—and love. Captivated by this image, I feel my heartbeat catch in my throat.

The vehicle coasts to a stop and Sean rolls down his window. He sticks his head out, "Buenas noches, Miguel, cómo estás? Necesitamos llegar al estanque."

Before I can muster a complaint, Sean lowers my side window. "Wait a minute, you speak Spanish? And how do you know Miguel?"

I watch Miguel and his son walk toward us from the barn. How could a man who'd been in town such a short time know so many people? And where has my head been? When had I stopped enjoying the beauty of the simple things around me? But the answer is easy. I've been consumed with my grievances, feeling sorry for myself. I'd wager if Prof. Fontenoy was still alive, he would trade anything to experience one more sunset with his loved ones. The thought makes me profoundly sad. I feel a deep sense of urgency to solve his murder.

"El Sherif," Miguel calls out.

I notice Miguel's son, Manuel, carries a small black and white puppy. His younger sister, Maria, follows closely behind. "Oh look, they've got a puppy," I start to open the door but realize the safety latches are on. "Hey, Sean, let me out?"

"Why don't we take care of our business at the pond? Then if we have the time before the council meeting starts, we can stop by and see the puppy," Sean says.

"You're right. Stripes has been stuck in this box for a while now. I'll explain to Miguel why I am showing up at his front door for the second time today. With the police."

"Oh, don't worry," Sean says. "Miguel and I are old friends. He's been helping me improve my rusty Spanish."

Sean launches into what sounds to my ears like a very fluent explanation of our mission in Spanish. I watch the expression on Miguel's face change from welcoming to one of concern and feel obliged to reassure him.

"The snake is perfectly harmless, Miguel. Besides, I'm sure you will never see him again after his misadventure today."

Miguel replies, "It's not the snake that concerns me, Señorita Indie. It seems you continue to have more than your share of trouble lately. You must be more careful. It is good you have such an ally as el sherif."

Trouble, there was that word again. And Sean as an ally? An intriguing idea, but not likely. Especially when he realizes I intend to meddle in his case again.

"That's a cute puppy you have there, Manuel. What's his name?"

"Her name is Margarita. She is a girl like me," Maria pipes up.

"It means Daisy in English," Manuel adds. "We chose a girl so that she can have puppies one day. Then I get to name the next one."

"That's a beautiful name." I look toward Miguel. "When do the sheep arrive?"

"Day after tomorrow. We are preparing the pens to separate them from Satin and Lace for a few days."

"That is probably a good idea, especially Satin. She can be pretty mean."

"Is Satin the llama?" Sean asks, shaking his head. "She spat at me when I tried to pet her. The other one's a sweetheart, though. Whatever possessed you to adopt those two?"

Both Manuel and Maria giggle, and I can't help but smile.

"They belong to my mother, so I've just been tasked to care for them." Along with every other creature in our ever-growing menagerie. "Which reminds me—we should probably get Bob to his new habitat."

"Bob?" Sean asks. "I thought his name was Stripes?"

"It was, but I decided to change it to Bob. I think a biologist like Prof. Fontenoy would appreciate the irony."

"Can't argue with that," Sean says.

We wave at the Gutierrez family and continue our trek toward the pond. It's a beautiful evening. As the sun wanes, the humidity finally dissipates. Sean leaves the windows down, and as the evening breeze hits my face, I notice an appreciable drop in temperature. I feel Bob move restlessly inside the box. He, too, senses freedom is close at hand.

"Which type of habitat do milk snakes prefer? The pond or the woods?"

"Wha—what? Oh. Milk snakes are adaptable to most environments, but the wood line would probably be best. Plenty of hiding places for Bob, here."

"How do you know Bob is a he?"

"His tail—its longer and thicker and tapers off suddenly. Females' tails are thin with a gradual taper. And judging by his size, he looks to be a juvenile. He'll probably top out at four feet, fully grown." I lean forward to gesture over Sean's shoulder. "Just park in front of the garage, and we can walk to the backside of the pond."

Sean swerves the SUV into a sudden arc and reverses back into the parking space. Another one of those police-isms you always hear about. He hops out of the vehicle and opens my door with a gallant gesture.

"After you, snake wrangler," he says.

"Hardly." The sun sinks below the tree line as dusk settles in. Once again, I feel Bob move inside the box. I hold the box away from my body, almost as if it could bite. Insert concerned face emoji here. "Perhaps you would like to do the honors?"

"Sure," Sean replies, a quizzical look crosses his face.

Taking the box, Sean walks beside me as we skirt the pond. The two beehives are visible from here, and I watch a few stragglers make their way back into the front entrance of the honeysuckle hive. The other hive remains conspicuously silent. I notice Sean glance toward them. "I came out to check on them earlier today."

"I trust you didn't find another dead body?"

"What? No, of course not. But one of the hives had absconded, and I'm unsure why."

"Absconded? Isn't that when bees abandon a hive for some unknown reason?"

I look at Sean with newfound respect, "Look at you, all knowledgeable about beekeeping."

"We do have apiaries in Nebraska. I'm from Elkhorn, Nebraska, if you didn't already know."

"I might have heard something to that effect." I smile at the self-satisfied grin on his face.

"Anyways, I've been doing some reading on beekeeping. I figure it can't hurt to become more educated. Who knows, it might give me the edge to solve this case."

"Wow, I'm impressed. You're taking this investigation seriously."

"That's my job. I'm here not only to find the murderer but also to try and get justice for the victim. Gathering information on what made Bob Fontenoy tick is just one aspect of determining who killed him."

"Oh yeah—so what made Prof. Fontenoy tick?" Not to be Captain Obvious here, but I wanted an edge too.

The mic on Sean's left shoulder squawks out of the blue, and we both start.

"601, what's your 10-20?"

Sean hands over the box and reaches to cue his mic. I feel Bob start to thrash about. It's no surprise that all of this erratic movement agitates him. Feeling queasy, I realize I'm no longer talking about the snake. I feel the urge to drop the box and run.

"This is 601, still at Evan's Road with 10-11," Sean says.

"10-4, 601. 580 requesting backup at 410 Main for 10-15."

"What's wrong?" Although I'm not current on scanner codes, I recognize the meaning behind requesting backup. My gut begins to churn in earnest.

"We need to release that snake and be quick about it. There is a civil disturbance brewing down at the Chamber meeting. Would you know anything about that, Indigo Bleu?" Sean asks.

Chapter Eleven

There was no question of me riding up front on the return trip to town. The slam of the SUV's heavy back door, along with the distinctive click of safety locks, has me feeling more suspect than—what? Sean Riordan and I had been having a moment. Hadn't we? But, whatever the moment might have become, it evaporated like a raindrop on hot pavement with the numbers 10-15, which I now understand means civil disobedience.

Sean drives faster than I would attempt while navigating the back roads. There is no witty rapport or wry smiles but rather an air of reflective silence. When I catch his eye in the mirror's reflection, he immediately looks away. "Sean—I just want to say—"

"Not now, Indie. I need to concentrate on the road."

The pounding of drums is the first thing I hear, followed closely by a loud man screaming into a bullhorn. "Get back! Back up!" I recognize the voice of Chief Deputy Harlan Pierce. *Uh-oh.*

"What the—" Sean asks. "Who are all these people?"

"I have no idea." But, unfortunately, I do. Heather's little bird has returned with its flock. The four-hundred block of Main Street is teeming with protestors. Since it's the main artery through town and only eight blocks long—well, you see the problem.

The Kanza Activist Coalition is a climate group loosely comprised of members of indigenous tribes, traditional inhabitants of the state. Based out of Lawrence, Kansas, the group fights for climate justice and water preservation. Both are noble causes, but an unwelcome disruption in a small town the size of Colette. The good folks of Corley County don't take kindly to outside interlopers. Sean utters an expletive, and I wisely keep my mouth shut.

Whup, whup. Sean taps the siren and utters an expletive. Deputy Hernandez swings back a wooden barricade. Sean maneuvers the department's SUV into a parking space close to the Chamber entrance and powers up the windows, muffling the drumbeat.

"Stay here," Sean says. He opens the driver's side door, leaving the engine and air conditioner running.

"Hey, you can't leave me locked up in here. I was planning on attending the meeting."

Sean sighs heavily, turns off the engine, then swings both legs out of the vehicle. He appears to stand there and take in the scene for a moment. Finally, he opened my door.

"Dressed like that?" Sean raises an eyebrow in query.

I glance down at my faded green *Colette Saints* T-shirt and jeans. Okay, not my outfit of choice to attend what appears to be a contentious Chamber of Commerce meeting, but at this point, I'm committed. "Is there something wrong with my fashion choice?"

Despite the noisy crowds, the challenge in my tone refocuses Sean's solitary blue eye on me. His steely gaze has me feeling as disheveled as I probably look.

"Nope," Sean says. He slams his Stetson onto his head.

Harlan Pierce, bullhorn in hand and visibly agitated, appears next to Sean. Streams of sweat run down his forehead.

"Riordan," Harlan says, his words clipped." I've cordoned off the perimeter and tried to push them back into a protest zone, but they keep sneaking past. A couple of them made it inside already. Should I arrest them?"

"No, Harlan," Sean says. "We still adhere to something called the First Amendment. So, as long as the protestors remain peaceful, no one gets arrested. Do you hear? But we limit the number of attendees inside per the fire code. You'll need to ensure the members of the council and designated speakers have seats. Do you have an idea of how many people are already inside?"

"Good thinking. I'll go run and get a head count." Harlan trots off, his duty belt flops about his waist.

Sean swivels his head to scan the burgeoning crowd. The drumming intensifies as a small group assembles behind the barricades. They hold placards with the skull and crossbones symbol, a universal sign of poison, crossed out with red paint. The vivid color resembles dripping blood. *Ew.*

The crowd chants, *"Hey, hey, ho ho. Climate change has got to go."*

Sean turns back to me. "Things have gotten interesting around here since I've made your acquaintance, Indigo Evans."

"Always happy to be of service, Interim Sheriff Riordan."

"I didn't mean that as a compliment," Sean says. "If you plan on attending the meeting tonight, I suggest you find a seat before Harlan finishes his count." He strides toward the temporary barrier. "Hernandez, push back the perimeter to street width and notify County EMS to be on standby."

I take this as my hint to leave and slip through the locals milling about. I won't argue with Chief Deputy Pierce about claiming a seat. Instead, I quickly text Heather to let me through the back door. At the very least, I can help serve cookies.

"Can you believe how many people showed up to protest?" Heather asks, her eyes wide with excitement.

"Heather, I'm not sure this was a good idea. Sean seems pretty pissed off." I transfer a platter of Angela's cookies to a table next to the coffee urn.

"Too bad! The Council shouldn't be allowed to cram through Agro-Tech's agenda without some form of pushback," Heather says. She sets out a stack of recycled paper coffee cups alongside individual honey straws and pitchers of almond milk.

"Can you point out Carson Wells to me?"

"Yeah, he is the slim, bald guy up front." Heather points to three men clustered toward the front of the room.

I scan the group. The two older men are not local nor, would I wager, protestors. With the heat building up in the enclosed space, the pair have abandoned their suit jackets. The younger man, whom I identify as Carson Wells, wears the same black monogrammed polo as Josh. His bald head is shaved slick, and he sports stylish, rimless glasses. He has the trim, toned physique of a runner. "Well, he's not exactly what I expected."

Heather shrugs, "He looks like a chemistry nerd to me."

Having spent most of my academic career thus far among nerds, I could see that. Anyone of import in our little town is gathered in the crowded room. I spot Dr. Victoria Medford, wearing conservative business attire, shaking hands with one of the Agro-Tech executives. She sits next to—wait a minute—is that Alex Carmichael? I recognize the back of his blonde head and the newly acquired linen jacket as he leans to one side to speak into Dr. Medford's ear. That rat! I duck my head when Alex turns to look at the crowd behind him. I don't want him to see me, especially in my filthy snake-wrangling clothes.

Frank and Brenda Matthews arrive. Brenda wears a pale blue silk wrap dress. She heads toward the line of chairs behind the speaker's podium on the small dais. Choosing the seat second from the right, she sits down. Mayor Frank sports an open-collared dress shirt and summer suit coat. He stops to glad-hand with many of the locals already seated.

Frank and Brenda are an enigma to me. Having migrated from Wichita, the pair set up shop in Colette four years ago. Determined to make a name for themselves, Brenda ran for various positions on civic boards and Frank in local elections. When Frank was appointed mayor last year, Brenda became the face of the real estate and rental business. Frank, a natural politician, verbalizes the correct platitudes of the impartial city official, but he spends most of his time advocating for business opportunities that enhance his own. I guess for a salary of less than ten thousand per year, I can't blame him. However, he does seem to travel a great deal, so much so the locals speculate he has a piece on the side.

Frank's effusive, over-the-top personality is fully displayed as he approaches the front. If chanting protestors bothered him, you would never know it by the back-slapping persona he exudes. Meanwhile, Brenda fusses busily with the V-neck of her dress. It gapes open, revealing plenty of cleavage, reminding me again of the note on my dad's cabin door: *"Let's talk."* What was that about?

Suddenly, I recall Brenda's impatient actions at the four-way intersection yesterday. I had assumed she was late for a yoga class, but with the murder of Prof. Fontenoy, the circumstances of the past twenty-four hours have drastically changed. So, the question foremost in my mind is whether Brenda Matthews had been in a hurry to get away from somewhere. Like a murder scene?

Frank moves behind the lectern and taps the mic. The reverb of the PA system stills the chatter in the room. "Welcome and good evening, everyone. We have a long agenda and, as you can see, a full house. So, grab some refreshments and find your seats. And, speaking of refreshments, I would be remiss if I didn't thank two of our local business owners from Angela's Sweets and the Bee Queen. Angela and Heather, thank you for your generous donation this evening."

Heather and Angela wave from behind the table to a smattering of applause. Trying to be inconspicuous, I press my back against the wall. My wallflower act must not work as someone yells, "Go Saints!" I look up to see Josh Blake waving enthusiastically from the side of Carson Wells and return the gesture half-heartedly. The two corporate men and Josh take their seats in the front row. Carson Wells bends forward to greet Brenda, getting an up close and personal view of her bosom. He sits in the empty seat beside her and surveys the chattering audience with a bland expression.

"This Colette Chamber of Commerce meeting is called to order." Frank bangs the gavel on top of the lectern twice.

I scan the crowd for possible demonstrators who might have escaped our eagle-eyed chief deputy and note a few who might qualify. Finally, Sean Riordan enters the room and stands with his back to the wall near the entrance. Deputies Nolan and Hernandez flank him. An overt show of force?

Mayor Frank muddles through the obligatory chamber announcements, loosely following *Robert's Rules of Order*. He determines a quorum of council members is present. No surprise there. This meeting is the hottest ticket in town. I laser-focus on a pair of young women I don't recognize. Could they be protesters? The tall one with long, dark hair falling past her shoulders wears a pair of Daisy Duke cutoffs and Western-style boots. The outfit accentuates her lean, tan legs. Beside her is a petite blonde dressed in a short crop top and tight white skirt. The blonde shifts nervously in her seat, exposing a sparkly navel piercing on a smooth expanse of pale skin. Oddly enough, both girls wear buttoned-up denim jackets in this overheated room. But the girls are pretty, so it's no surprise they've managed to charm our not-so-diligent chief deputy and claim a seat.

I glance back to find Sean studying me and try to draw his attention toward this pair of potential troublemakers. In what I assume to be a conspicuous manner, I switch my gaze back and forth. Sean leans forward and whispers in Deputy Nolan's ear. The officer steps out of the room. What the—?

"It's my honor to introduce Carson Wells, the county's plant manager for Agro-Tech Industries. Carson is going to give us an overview of the expansion plans for the Colette division, which will bring an additional two hundred well-paying jobs to our area," Frank says.

A flurry of whistles and cheers accompany this announcement.

From the back row, a young man with a single braid of long black hair leaps to his feet and yells, "Filthy jobs that poison our land and water. What do we do when Mother Earth is under attack?"

Two other men seated nearby jump up and raise their fists. They begin to chant, "We stand up and fight back! We stand up and fight back! We stand up—"

Harlan Pierce grabs the demonstrator nearest to the aisle and drags him out by the collar. The young man promptly goes limp, becoming dead weight in Harlan's firm grip, but his slender build is no match for the deputy's beefy stature. Harlan hauls him up one-handed by his waist and carts him out the door. All the while, the protestor shouts, "Stand up and fight back!"

Deputy Hernandez quickly escorts the young man's companions out of the room. Both have raised fists and shout, "Mother Earth is under attack." The reply from the second is drowned out by the locals cheering and clapping in response to the Sheriff's Department's decisive actions.

Mayor Frank continues to extol the virtues of the Agro-Tech expansion as Carson Wells fidgets with his laptop and projection system placed on a table next to the lectern. I keep my eyes on the two girls I assumed were protestors. Thus far, they've been well-behaved, showing no more than idle curiosity in the trio escorted from the room. Maybe I was wrong about them.

A graphic appears on the projection screen: *Agro-Tech Industries Projected Growth and Revenue Analysis.* The accompanying graph identifies current statistics as well as projected growth charts. The county's potential revenue stream diagram is particularly interesting to the crowd. And yet another slide shows the gross salaries the additional labor force will add to the county's tax base. The crowd cheers at the salary projections highlighted in red.

Carson Wells clears his throat and begins to speak, "As you can see by the cost-benefit analysis, if the Chamber approves our expansion, it will—"

A collective gasp comes from the audience. Carson freezes mid-sentence. The long-haired cowgirl has climbed atop her chair. She rips off her denim jacket, revealing naked breasts painted bright red. She yells, "Corporate scum, you pollute and exploit the planet! The land is not for profit."

At this point, we could have sold tickets to the event and raised a ton of money for charity. But I have to give the group kudos for their ingenuity. The audience turns to gawk at the young woman, utterly naked from the waist up. With her fist raised high, she resembles a member of Amazonian royalty shouting about injustice and corporate greed. Preach it, sister!

I notice her companion slip into the aisle and dart toward the lectern. She, too, has shed her jacket, revealing red-painted breasts. However, hers are not nearly as impressive as her cowgirl friend. Just saying. Still, she holds an opaque plastic bag in her hands, and it looks a lot like—wait a minute—is that a blood bag?

From this point on, everything happens in slow motion. Nolan and Hernandez reach the warrior princess's chair and try to pull her from it. Unwilling to go down without a fight, she slaps at their hands, screaming about police brutality and assault. Nonplussed, the deputies drop their hands to their sides, looking bewildered as to what to do next.

Not to be outdone, the little blonde almost reaches Carson Wells before he notices her. She pulls the plug from the top of the blood bag and starts to squeeze. I must say, for a chemistry nerd, Wells has ninja-like reflexes. He manages to duck aside and scoot his laptop from the blood's trajectory. A stream of red liquid, which I hope is fake, spurts out and hits Brenda Matthews square in the chest. It runs down the V of her neckline and soaks the silky material. A subsequent squirt from the red-breasted pixie hits Brenda square in the face. Brenda stands there stunned, looking like the horror movie version of an aging prom queen. I don't know whether to laugh or cry.

Out of nowhere, Sean Riordan closes in on the blonde. He slaps the bag from the girl's hands before she can do additional damage. The contents splatter across the legs of his jeans, and the bag lands flat on the floor—the residual hemorrhages onto the dais into a puddle of bright red.

I glance over to see Heather and Angela laughing hysterically. Who knew Chamber of Commerce meetings could be so much fun?

"We should probably start packing up, ladies. I'm pretty sure this meeting is over." It's a struggle to keep the grin off my face.

Chapter Twelve

- -

"Did you see Brenda's face when the fake blood hit her? I thought I was going to pee my pants I was laughing so hard," Heather says. "At least, I think it was fake. Don't you, Indie? I mean—where would they get the real stuff?"

"I'm sure it was fake." I yawn and feel my jaw pop. Ouch. Between yesterday's pointless sleuthing and the excitement of last night, I hadn't slept well. Again. At least lying awake until the early hours had kept me from reliving more bad dreams. But this morning, Heather's constant chatter and insistence I help restock the shelves in the shop gave me a killer headache. Ugh, I didn't mean to say killer. Wait—did I say that out loud?

"Indie," Heather says. "Are you listening to me?"

"Wha—what? Yes, of course. I have a headache. Do you have any magic brew that might help?"

"Indie," Heather laughs, "you may poo-poo my tisanes all you want, but I have a peppermint tea guaranteed to ease your headache. It's a completely natural pick-me-up as well."

"Sold." The way I feel now, I'm game for anything. So, awaiting my magic cure-all, I climb the stepladder to restock the upper shelves with Karl's honey… *ahem*… correction: Karl's and my honey. I take time to appreciate the sight of the bottles filled with liquid gold. They remind me I need to check on Karl. After all, he's well past his prime, and finding a dead body among his beloved bees must have been a shock.

"I'll put the kettle on," Heather says. "Be right back." She pushes through the swinging doors into the commercial kitchen area.

The doors shut, blocking the din of Angela's metal cookie trays sliding onto cooling racks. Ah—peace at last. I close my eyes and begin to massage my temples.
"Indigo?"

My mother's quizzical tone disturbs my moment of zen. It's an unpleasant reminder she's been conspicuously absent the past twenty-four hours.

"Indie! Why are you standing on a stepladder with your eyes closed? You could have fallen."

Like that, the ever-revolving door of mother-daughter relationships turns again in her favor. "Good morning to you too, Mother." I pause mid-step as Claire, resplendent in a vibrant red tunic and wide-legged, black-and-white striped pants, steps into view. From my vantage point, she resembles a barber's pole. But if anyone can carry off the look, it would be Claire. My mother's the ultimate queen-ager and wears the crown well.

"*Cara*, don't frighten her so." I hear the heavily accented voice of a man from behind Claire. He steps forward, and I glimpse a man who can only be our university's new artist in residence, Alberto Giovanni Vitale. As for the list of hot men in their forties, he would rank near the top. Supposing I kept such a list, of course. Tall and muscular, Giovanni has dark hair that falls in soft, loose curls brushing his broad shoulders. I would liken his angular profile and dark brown eyes to the face of a fallen angel. A divine messenger who has somehow managed to land in the middle of flyover country. Hallelujah.

"Let me assist you, *per favore*."

I'm a sucker for a man with an accent, and *Signore* Vitale's voice is delightful. Without hesitation, I place my hand in his and allow him to assist me down the stepladder. I slowly descend, feeling as delicate as a porcelain doll.

"Indigo, this is Giovanni Vitale. Giovanni, meet my eldest daughter, Indigo Bleu," Claire says.

Do I detect a note of disapproval in Claire's tone? Yippee.

"Indigo Bleu—*cara*," Giovanni exclaims. "A suitably beautiful name."

91

My hand remains within his firm sculptor's clasp. Giovanni leans forward to kiss my cheeks: European style and swoon-worthy.

"Mother." Heather pushes through the door holding a tray loaded with a china teapot and cups. "Where have you been hiding?"

Is that a note of accusation in Heather's tone I hear?

"Whatever do you mean, Heather," Claire titters. "I told you I would be helping Giovanni settle in."

And no, I did not misspeak using the word "titters." It's not a word I associate with my mother, yet there it is. I release Giovanni's hand and give him a grateful nod. He smiles, revealing a mouth full of pearl-white teeth. I swear I hear the angels sing. Then he moves toward Heather, his hands outstretched, reaching for the tray.

"Heather. How can it be, *cara*, that you have two such beautiful daughters?" Giovanni asks.

Intrigued, I watch Heather. She frowns and hangs onto the tray a bit too long. Wyatt steps through the door and almost bumps into her backside. Where did he come from?

Finally, Heather relinquishes the tray and turns to Wyatt. "Can you put the kettle back on, Wyatt? And we'll need more cups. Or perhaps you would prefer an espresso, Mr. Vitale?"

"Oh—*si, cara*. An espresso would be *magnifico*. And please call me Giovanni. Mr. Vitale sounds too much like *mio padre*," Giovanni says.

"Pfft," Heather whispers.

Wyatt takes in Giovanni's overwhelming presence in our little melodrama, but then the younger man turns obediently to do Heather's bidding. Poor kid, I need to have a talk with him about her.

"I was very young when I had my girls," Claire insists. "Heather, I'll have an espresso too. But only one. We need to get down to the art department this morning."

My mother and espresso? A toxic combination if there ever was one. Heather must have read my thoughts, and I hear her say, "Perhaps, you might prefer a cup of peppermint tea instead. It will give you a gentle form of energy without all the jitters." She points at Giovanni.

"Ah, your daughter is correct, *cara*," Giovanni says. "Normally, in the morning I would drink cappuccino, but with the jet lag—an expresso will do. I am still adjusting to the time change." He expressively shrugs his shoulders.

"Oh, very well, I suppose the tea will do," Claire says. "So, what did I miss yesterday? Have they found out who killed poor Bob?"

"Funny, you should ask," Heather says. "And no, we haven't found out who killed Professor Fontenoy. Yet. So, again, just where were you all day, Mother?"

I've never heard Heather use such an accusatory tone with our mother. Suddenly, I'm living the Midwestern version of *The Parent Trap* and switched personas with my sister. No way am I prepared to let that happen. Reclaiming my status as the contrary daughter, I speak up. "Well, Claire... to answer your question, yesterday, a snake was loose in our apartment, and a protest got out of hand at the Chamber meeting. But, other than that, no leads in the case to speak of."

"A snake!" Claire shudders. "What in the world—? And Heather, when you said "we," what exactly did you mean?

"*Oddio!*" Giovanni exclaims. "So, this is the wild west I've heard so much about. Do you think I will be able to see a—what do you call it—a grizzly?"

"Yes, but there are no grizzly bears in Kansas." I hasten to reassure Giovanni. "Black bears live in southern Missouri. We only have opossums, raccoons, and coyotes around here."

"Really? Perhaps I will incorporate several wild American creatures into my first sculpture," Giovanni says.

"The point is—" Heather interrupts, "—a snake was set loose in our apartment, and Indie had to wrangle it out. And when I say 'we,' I mean Indie and I are trying to solve Professor Fontenoy's murder to clear her name."

"Don't forget me!" Wyatt steps through the door holding the tray with a small cup of espresso and extra teacups. "I'm representing Indie."

"*Contestatori?*" Giovanni says. "*Cara*, you told me nothing exciting ever happens in this *piccolo città*."

"Yes, ah—" Claire says, her voice vague. "I did hear some gossip about protests getting out of hand last night."

"Oh yeah—what exactly did you hear?"

Oh boy, here we go again. How is it that Sean Riordan has the uncanny ability to show up at the most inopportune moments? And how can he be so stealthy for such a tall, well-built man? I look pointedly at Sean. "We're not open yet."

"Really?" Sean says. "The door was standing wide open."

The arrival of another testosterone-laden male snaps Claire out of her malaise. "You must be the new sheriff. I've heard so much about you. All good, of course," she adds. I'm Claire Evans, Heather and Indie's mother. What are you doing to get Bob Fontenoy's murderer off the streets before he hurts someone like my daughter?" Claire asks, a slight edge to her voice. She stretches her hand forward, fingers drooping languidly.

The edge of disdain in her tone is not lost on Sean. He glances toward me, and his jaw tightens. Sean doesn't know whether to shake or kiss Claire's hand. Finally, he compromises, holding the tips of her fingers while doffing his Stetson.

"A pleasure to meet you, ma'am," Sean says. "I'm just the acting sheriff. And I'm not at liberty to discuss the details of the investigation."

Well, color me officially impressed. Sean directs none of the usual flattering comments toward Claire, such as "I can tell where your daughters get their good looks" or "You look as if you could be their sister." Nope, Sean Riordan states the facts politely but firmly. I feel my heart start to warm toward him.

"And that includes the two of you." Sean looks toward Heather and Wyatt.

Then he levels a blistering glare at me, and my warm feelings blaze red hot. Not in a particularly affectionate way.

"And especially you, Indigo. I agree with your mother. You need to stand down until we determine the killer's identity. For your own safety."

"Well, maybe, instead of bugging me, you should be out doing your job—"

"Well, maybe if I didn't have to worry about you getting hurt," Sean retorts.

"*Impressionante!*" Giovanni interrupts. "A real American lawman. Is that a cowboy hat you are wearing? Where can I get one? And what kind of pistol is that? Will I be able to shoot a gun while I am here?"

Distracted by this barrage of questioning, Sean pivots. "And you are?"

"*Permettimi*, I am Giovanni Vitale of Milan, Italy."

Maybe the expression on Sean's face and the hand on the butt of his service weapon keep the traditional Italian greeting of two kisses on the cheeks at bay.

Giovanni takes a cautious step back. Ding, ding, ding. Our intrepid lawman wins round one of the toxic masculinity game.

"Well, Mr. Vitale, one has to be licensed and properly permitted to carry a firearm in Kansas," Sean says. "But I'll be happy to recommend a place to buy one of these." He tips the brim of his hat. "Just make sure you buy an authentic *Stetson*. They last forever."

Studying Sean with the discerning eye of an artist, Giovanni asks, "Perhaps you would consider posing for me? Your profile begs to be chiseled in stone."

Whoa. Italy just scored on a penalty kick. Everyone is riveted in place, silently assessing the other's reaction. Heather giggles, breaking the silence. My mother cracks a smile. Poor Wyatt observes the back and forth between the two men, grateful he has yet to be challenged in this tree-marking display.

Giovanni's offer dangles tantalizingly in the air. I take pity on Sean. "Giovanni is the artist in residence for Colette University's Art Department. He is a sculptor, and he's Italian." As if that means anything. "Was there some other reason you came by this morning? Other than to yell at me?"

"I see," Sean says. "I'm not here to yell. Yes, there is a reason I'm here this early. I wanted to speak with Ms. Heather about last night's incident."

Busted. I look at my younger sister and notice a slow flush of red creep up her neckline. Guilt? Hate to say it, but I told you so.

"I had no idea those girls were going to strip and throw red paint on Brenda Matthews," Heather blurts out. "It wasn't actual blood, was it?

"It was red food coloring," Sean says. "But Brenda's dress is ruined."

"Brenda Matthews? What in the world does that woman have to do with anything?" Claire asks.

"One of the protestors mistakenly hit Brenda with blood at the Chamber meeting last night—or—eh, I guess it was red dye in a blood bag. Carson Wells, the plant manager at Agro-Tech, ducked and Brenda got splattered. It was running down her hair and onto the front of her dress." I try my best not to grin at the memory.

"You forgot to tell her the protestor was naked from the waist up and had her boobs painted red," Heather adds.

"It was epic," Wyatt asserts.

Epic? I look at Wyatt with newfound respect, and I have to agree. It was epic and the most excitement this town has seen since Prof. Fontenoy was found dead in my bee yard some thirty-six hours ago. Like that, a hypothetical bucket of cold water dumps on my head. Focus, Indie. Do not allow distractions, including whatever is between you and Sean Riordan, to interfere with your primary goal of discovering who killed Prof. Fontenoy.

Abruptly, I change the subject. "What do you know about Brenda Matthews, Mother?"

"Well, hardly anything other than Brenda is quite persistent in her job and willing to do almost anything to accomplish it," Claire says. "And I do mean anything."

"Ah—*alla puttanesca.*" Giovanni says.

"Oh, I love that dish," Wyatt murmurs.

"No, Wyatt, I believe Giovanni was insinuating Brenda Matthews figuratively prostitutes herself to get the job done. Right?" Heather asks.

"*Si, cara.* Perhaps, I do not personally know this woman, Brenda. But I have met many like her," Giovanni says.

With Giovanni's looks, I bet he has. Glancing at Sean, I note he seems to be considering Giovanni's words.

"Oh," Wyatt says. "I thought he was talking about the pasta sauce." Wyatt blushes furiously. "Still, it is one of my favorites."

By now, I'm in for a penny and much more than in for a pound. Since this is one conversation I prefer not to have with my father, I ask, "Mother, what do you know about Brenda Matthews and the farm?"

"The farm!" Claire exclaims. "What did that woman say about the farm?"

Whoa—I've struck a nerve. "It's not what Brenda *said*, Mother. I was out there yesterday and found one of her real estate flyers pinned onto the door of Dad's cabin with a handwritten note on the back. Would you know anything about that? Is Dad selling the farm?"

"No, not the farm," Heather cries out.

"No, Indie, of course not," Claire insists. "Your father would never sell the farm. And if Brenda Matthews thinks I'm going to let that happen... Unlike your Aunt

Rachael, your father agrees the farm is part of your girls' heritage and should remain in the family."

I must admit to feeling relieved at my mother's words, but that doesn't explain Brenda's presence or Claire's apparent antipathy toward her. Looking at my mother's face, I decide she's had enough questions. Good grief, I must be getting soft. "Okay. We can discuss it when Dad returns."

Relief floods Claire's face, and Heather bites down on her lip. My sister wants to say more, but airing family laundry is not an Evans trait. More's the pity.

Of the three men present, Giovanni watches us with avid curiosity. Meanwhile, Sean stands by, silent as a stone. A mental image of what he might look like, sculpted in granite, naked, flits through my mind. Oy.

Poor Wyatt, recovering from his last faux pas, keeps his eyes downcast. "Wyatt, how about you and I run out to check on Karl? We can pick up some honey while we're there."

"Yeah—I mean—yes, I'd be happy to do that," Wyatt says, failing to hide his eagerness. "We can discuss your case."

"Sure, great." Yay for me.

Chapter Thirteen

- -

"**I**f we need to mount a defense, what direction should we take?" Wyatt asks.

"What—why would you ask such a thing? There have been no formal charges." A defense? I am starting to wonder about this kid's law acumen. I need to stop thinking of him as a kid. Sure, compared to the testosterone-laden Sheriff and sculptor pair we have left at The Hive, he was a kid, but— "How old are you, Wyatt?"

"Twenty-three," Wyatt responds automatically. "I turn twenty-four in August, and I realize that. I'm thinking hypothetically, gaming out various scenarios, trying to formulate a plan—just in case."

"Just in case what—?" The front right tire drops off the edge of the pavement as I overcorrect the steering wheel. Wyatt grabs for the overhead handle. Ha! Serves him right. "Just in case they charge me for murder because I found a dead man beside my beehives. Mind you, a man who was Chair of the Science Department at the university, I left under rather fractious circumstances, derailing my academic career?"

Okay, when you say the words out loud—perhaps we should have a plan, just in case. I am innocent and believe truth and justice will prevail, but who am I trying to kid? As the daughter of a lawyer, I know firsthand how easily a defendant might find themselves accused of far less. Perhaps I should drive off the road. "Hypothetically speaking, do you think there is a chance I will face charges?"

"No, not at all—eh—maybe? If something doesn't break in the case soon, they might devise some bogus charge to satisfy public demand for action. And from what I can tell, you are a viable suspect, ex parte as the evidence might be. I don't need to tell you how much pressure law enforcement is under to clear cases. The longer the killer remains at large, the longer the Sheriff's Department looks incompetent." Wyatt adds, "However, proving guilt at trial would be another thing."

"Trial? I plan to solve the case long before that happens." No way was I going to trial. "Who do you think did it, Wyatt?"

"If I had to make a guess, my money would be on Carson Wells," Wyatt says.

"Carson Wells?" Carson Wells was on my suspect list too, but down toward the bottom. "Why do you think it was him?"

"I don't know—I just have a feeling about him. I did some reading on the grant awarded by Agro-Tech Industries, and Wells has a lot riding on that new product line of theirs, ATI—Glo—?"

"ATI-EX 50/50. It retails under the trade name Glo-Grow. Why do you say that?"

"I have a buddy who manages the social media pages for the Agro-Tech plant, and he says Carson Wells was sent by corporate to ensure Glo-Grow scales up for national distribution. Wells is ambitious, and managing the Colette division is his stepping stone."

"Well, that doesn't mean he killed Professor Fontenoy." But, seriously, is there no one who is not on social media these days?

"It doesn't mean he didn't," Wyatts responds. "Maybe Fontenoy discovered discrepancies in your data versus the submitted grant proposal stats? They tout the study as a big part of their marketing campaign."

"True. Maybe—but wouldn't that implicate the proposal's author, Alex?" As much as I detest Alex Carmichael for his betrayal, I don't want to think of him as a murderer.

"Yeah, I guess, but Alex Carmichael seems a little too obvious, doesn't he?"

99

"Well, my money was on Phoebe Sutter. If it came down to it, there wasn't anything she wouldn't do to help Alex. Even if it meant bashing Professor Fontenoy over the head with a brick, Alex would probably take over his position within the Science Department. Couldn't that be construed as a motive?"

"I don't know. It still seems pretty far-fetched. Wait a minute—you said your money *was* on Phoebe?" Wyatt asks.

"Yeah, at first, I was positive she did it, especially after finding poor Bob curled up in our apartment yesterday."

"Bob?"

"The snake—you must keep up, Wyatt. Phoebe was in The Hive yesterday, and everyone knows she has a thing for snakes. She's got that snake tattoo crawling up her backside—"

"Phoebe Sutter has a snake tattoo on her butt? How do you know that?" Wyatt's face reddens. He clears his throat self-consciously. "Oh—I see."

"But, now, I'm not so certain it was her. You see, the thing is, Phoebe loves snakes. So much so that I can't see her putting a fine specimen of a Central Plains milk snake in danger—possibly killing the poor thing."

"I think Heather wishes you had killed it. She said she dreamed of snakes crawling up the walls all night long. So, if you don't think it was Phoebe, who was it?"

"How much do you know about Brenda Matthews?"

"Nothing. Other than Matthews Realty is the leasing agent for my apartment complex."

"Really? Interesting."

"Matthews manages most of the rental property around here, at least all that I know of. And she is the mayor's wife. What possible motive connection could she have to the murder of Bob Fontenoy?" Wyatt asks.

"I don't know, but that's what I'm trying to figure out." Matthews was the leasing agent for our building as well, but I'd left all the business details up to Heather, preferring the manual labor parts of the job.

Silence ensues as I contemplate the amount of circumstantial evidence *writ large*. I speak legalese myself. Ha! Who am I kidding? I speak law as well as I speak Spanish. Thank goodness Sheriff Kramer had the foresight to appoint Sean

Riordan on an interim basis. I'd be behind bars if Chief Deputy Harlan Pierce led the investigation.

"Did you speak with my father yet?"

"Yes, we spoke last night. He said he'd tried calling you, but you never answered," Wyatt says. "He doesn't think anything will come of it, but he said he will fly home immediately if any charges are applied."

Okay, either my father is overly confident in our local law enforcement or doesn't want to forgo his golfing time with his buddies. Either scenario has me feeling uneasy. "Wyatt, you know my sister is in a serious relationship, right?" I ask, changing the subject to something slightly more palatable.

"Yes, I know that," Wyatt says. "Why?"

"Well, I just thought—I don't want you to get hurt," I blurt out, trying my best to add a note of empathy. But subtle, I am not. "Heather and Austin have dated a long time."

"I know. Heather says they've known each other since grade school and started dating in high school. Austin has always dreamed of flying jets in the military, and Heather supports his plan."

"Okay, then—?"

"Look, Indie," Wyatt continues. "I'm fully aware Heather is off-limits. But she is funny and kind and has become a good friend. I am under no illusion she will drop First Lieutenant Austin Vale and fall for me. I swear I'm not some crazy stalker. I like helping out at The Hive. It's like being a member of one big family. Besides, Heather needs the help."

"I know that—I mean, I didn't mean to—" The words trail off. Is Wyatt somehow insinuating I wasn't helping Heather? Family? How well do I know the other vendors? I know Angela and Serena and now Bea, our clairvoyant senior citizen, and that guy with the *Rasta* braids makes the thingamabobs—I think his name is Jonathan. I know them, but certainly not well enough to consider them family. Okay, not true.? Once again, my self-absorption with personal woes has taken precedence over enjoying my life.

"You're right. I tend to allow Heather to carry most of the day-to-day management of the business. It's not my thing." Ouch—selfish I am, says me speaking Yoda-esque and not in the voice of the cute baby one.

101

"Heather knows you've been going through a rough patch," Wyatt says.

"That's hardly an excuse. Heather gets some much-needed time off after I check on Karl and pick up the honey. Starting today!"

"How long have you known Karl Bauman?" Wyatt asks.

"Most of my life. My family always purchased honey from him. Last fall, he helped me with my research project and mentored me as his apprentice this spring. Why do you ask?"

"Ah, no reason," Wyatt says. "I just wondered. Do you like him?"

"Yeah." Okay, that was an awkward segue. "You don't think Karl had anything to do with Fontenoy's death, do you? Because I can tell you right now, he did not. Besides being too old, he is as honest as the day is long." Karl, a contemporary of my grandfather, was from the school of thought that when the going gets tough, you dig deep and work harder. "He can be a little crusty sometimes and has little to say, but I've learned to listen 'cause it's always important when he does speak."

"My grandmother was like that," Wyatt says.

"Was?"

"Yeah, she passed three years ago. She practically raised me, and I still miss her. My dad was killed by a drunk driver when my mom was pregnant with me, so we moved in with my grandparents. My mom was—still is a nurse. She worked the night shift, so I spent most of my time with my grandparents."

"Oh, Wyatt, I am so sorry."

"It's okay. I never met my dad. I guess I didn't miss him, just the idea of him. But from everything I've heard, he was a good guy. My mom never remarried, so I'm an only child. Then, after my grandfather passed away, it was just me, my mom, and my grandmother. Probably why I get along with older women so well," Wyatt says with a self-deprecating laugh. "But, before he died, my grandfather taught me a few other handy things… like how to change a flat tire, check the oil, etc."

"Ha! My granddad did the same when I got my learner's permit. And he taught me the gas gauge wasn't the only important dial to watch. He could have never imagined the concept of self-driving cars."

Ruminating over my loved and lost grandparents almost makes me miss the turn to Karl's place. The gravel drive is narrow and lined with locust, box elder trees. The former is a favorite of bees in early spring, and the latter is a good source of sustenance for the birds and squirrels in late fall. Tall overhanging branches provide shade from the hot sun, already heating the exterior temperature.

"It's going to be a hot one." I remove my sunglasses to adjust to the sudden onslaught of shade and notice Gertie parked beneath the open bay of the detached garage. Good. The presence of the cantankerous truck means Karl is at home. I'd tried calling, but Karl was not tied to his phone. What must that be like?

"Cool truck," Wyatt exclaims. "What is it—like a '68 or '69?"

Of course, he would like Gertie. He did say he got along well with older women. "I believe it's a '69, and that truck is—"

Already enraptured, Wyatt asks, "Does it have a three on the column? My grandfather had one like that. He used to let me practice shifting the gears. Unfortunately, my grandmother hated driving it, so she sold it soon after he died. But I sure loved that truck."

"Oh yes, Gertie has the gearstick from hell." Knowing Wyatt is itching to look closer, I park behind the truck. "Karl usually keeps it unlocked so you can look inside. Be sure to take note of the lack of modern conveniences... like air conditioning. I hear Elvis howling, so Karl must be in the house. I'll go find him."

"Elvis? Gertie?"

"Elvis is Karl's basset hound, named after the king of rock and roll for obvious reasons. Gertie is the truck's name, apparently an homage to Grandma Gertrude." The dog's wail reaches a crescendoed peak. "That's strange. Karl says Elvis is better than any security system. I hope Karl is not sick—"

"Gertrude?" Wyatt asks. "Grandma Gertrude—" He repeats the words. "I'll take a look in the truck later. First, let's go find Karl."

Weird. Wyatt strides toward the house's front porch, almost like he is on a mission or has been here before. I hurry to catch up with him, and we take the front steps two at a time. Elvis's howls accelerate and change into a string of frantic barking. Now, I am genuinely alarmed. "Karl!" I bang on the wooden door. Elvis begins to whine and scratch behind it. "Karl! It's Indie. Are you in there?"

For the second time in the past two days, I turn a doorknob to someone else's house, but this time it's unlocked. All nose and ears, Elvis scampers out of the gap.

He whines and jumps on his short legs at my feet. I reach down to pat his head. He snuffles, his long ears flapping. "Calm down, boy. Where's Karl?"

Elvis yelps once, then takes off down the steps with his nose to the ground. "Karl," I call out through the opening. "Ew—" The pungent odor of doggie doo wafts through the doorway. "What? Karl, are you okay?"

"Should we go in?" Wyatt asks.

"Something is wrong. Karl would never leave Elvis unattended. We should check to see if he is sick or—worse." Pushing through the entryway, I carefully skirt the pile of dog poop. "Poor Elvis, he would never... Karl, where are you?" I feel a spurt of panic as my heart rate accelerates.

"Maybe I should call 911," Wyatt suggests.

"Maybe, but let's just check the house first." The house is an old two-story farmhouse, not unlike my grandfather's. A long hallway divides the small rooms on either side, and I know from experience the general layout. I've been inside the home a few times, usually to borrow the facilities. "Karl?"

I glance into the rooms, my voice sounding loud in the space. The house remains eerily silent as we traverse the hallway. The unmistakable tick of a German cuckoo clock hanging on the wall accompanies us. I hear the mechanical gears whirl, and a small wooden door at the top of the clock opens. A brightly painted bird perched on a lever pops out. *Cuckoo—cuckoo—* I count to ten, grateful it's not the strike of high noon this time. "He's not here."

"How do you know? We haven't even looked upstairs," Wyatt says.

"Elvis is howling outside. That dog may be getting old, but he always knows where Karl is. Come on, let's go check the workshop."

I take off at a dead run back down the hall, Wyatt close behind. We fly through the front door and barrel back down the steps. Now I hear Elvis baying the loud, confident cry of a scent hound who has treed his prey.

We round the corner of the house, and I see Elvis scratching frantically at the side door of Karl's workshop. The ubiquitous metal building sits off to the side of the house. It's the most modern structure on the property. I reach the side door and turn the handle. "Hold on, Elvis." But Elvis, hot on his master's trail, scrambles through the doorway and into the industrial space Karl uses to build hive bodies and repair frames.

Unlike my father's finished barndominium, this simple 24'x36' structure combines partially finished and unfinished workspaces. Karl completes the beehive woodwork in the front of the building behind the roll-up garage door. The big door is left open on hot summer days. Today, it's stifling hot inside the closed space.

I follow the sound of Elvis's feet padding through the stacks of partially completed brood chambers and skirt the wooden sawhorses piled high with lumber. "Karl—?" *Silence.* "Wyatt, open the door to let some light in. The latch is on the left side of the roll-up."

The door latch clicks, and a flood of sunlight illuminates the space. Immediately, I notice two things: Karl is not here, and Elvis is sniffing at the door to the honey room.

Karl's honey room is insulated to regulate the temperature for honey stability. Expensive extraction equipment fills the room. But, the gallons of liquid gold stored on the shelving units sum up most of Karl's life's work. For that reason, Karl keeps the door locked, and this time is no exception.

Elvis sniffs at my feet. He whines softly at the closed door. The hound is never allowed inside. Karl, a stickler for hygiene, considers dog dander a potential source of contamination. "It's okay, boy." Elvis looks up at me with large, soulful eyes. "Do you think he is in there?" I run my fingers along the narrow frame above the door.

"What are you doing?" Wyatt asks.

"The door's locked, but Karl keeps an extra key above the doorsill. I've got it—" I slide it down to the end of the frame. *Clink.* The metal key falls onto the concrete floor. Wyatt is the first to reach it, and he fumbles while placing the key into the lock.

"Clockwise."

"Got it," Wyatt says.

The door swings wide open. Elvis scoots between my feet and into the gloom. "Elvis, come back here." It's too late to worry about dog hair. "Karl?"

I grope for the light switch as the sunlight streams behind me, but again, Elvis's nose saves the day. He erupts in a series of short, excited yelps signaling the objective achieved. As the fluorescent lighting floods the space, I hear a low groan from the back of the room. "Karl! Oh my God! Wyatt—call 911!"

Chapter Fourteen

"**W**hat else do you want me to say? We found Elvis shut up in the house when we arrived." Once again, I describe the sequence of events to Sean Riordan as we watch County EMS load Karl's stretcher into the back of the ambulance. "We followed the dog to the workshop and found Karl locked in the honey room. We called 911 immediately."

"Elvis—now there's an appropriate name for a basset hound," Sean murmurs.

We both look to where Chief Deputy Harlan Pierce is questioning Wyatt. They stand beneath the shade of a large oak tree while Elvis lolls on the ground at Wyatt's feet. The dog's tongue is out, and he pants heavily. Though having never met Wyatt before today, Elvis follows him everywhere.

"Do you trust him?" Sean asks, watching the pair with his deputy.

"Who? Elvis? Sure."

"Not funny—you know who I mean. You come out here alone with some kid, barely wet behind the ears, and the two of you find an old man who is probably concussed and locked in what appears to be a storm shelter. Christ, Indie, what were you thinking?" Sean asks. He combs his fingers through his hair, frustration evident. "Just how well do you know this Wyatt K. Price?"

"I'm sorry. I tend to make stupid jokes when I'm nervous." Distracted, I force myself to stop fantasizing about running my fingers through Sean's thick, dark hair. Bad girl, Indie.

"I've no reason not to trust Wyatt. I don't know him as well as Heather, but he's been a huge help at The Hive. And besides—Elvis likes him, so there is that. I think my father likes him. Why else would he tell Wyatt to represent me while he is out of town?"

"You think, or you know? Is it just me, or am I the only one to notice Price is conveniently standing by whenever things go sideways around here?"

"You could say the same about me."

"Oh, don't get me started—at least half of the Sheriff's Department thinks you should be locked up."

Because they think I killed Professor Fontenoy?" Surprised, I feel genuinely affronted by this admission, but I suppose I deserve it.

"Well, only a couple think you're guilty of a crime. The others want you locked up for your own safety. I'm half inclined to agree with the latter," Sean says.

"Oh, do you?" I glance toward Chief Deputy Pierce and tilt my head to the side. "I'm sure I know one officer who would find me guilty, but who are the others?"

"Never mind, I shouldn't have mentioned it. I'm sure it's no revelation to you. Our Chief Deputy has made his feelings abundantly clear. By the way, I meant to ask, what's up with you and Pierce?"

"Oh, he never got over me nailing him in the groin with a dodgeball in the seventh grade."

Sean coughs and then winces. "Nailing him—what? Ouch! Why did you do that?"

"I told him not to call me DB, but he did it anyway."

"Why DB? Oh, I get it, Indigo Bleu—double blue," Sean laughs. "That's not so bad,"

"Oh, yes, it was. I was one of those tall, skinny girls who towered over all the boys in junior high. Harlan was always teasing me about my height. He thought it would be cute to give me a nickname. It finally dawned on him that indigo and blue were the same color, and he kept taunting me about it in gym class. One day, I'd finally had enough and kicked the ball straight at him. It hit him point-blank in the groin. The way he carried on you would have thought I'd killed him. The gym teacher escorted him to the school nurse."

Sean laughs. "I can just see that. You know seventh-grade boys are generally pretty cruel?"

"Yes, I know. But Harlan failed to realize tall seventh-grade girls kick far and hard. He was humiliated and has never missed an opportunity to call me DB."

I notice Sean grimace and, like a typical man, reach down to adjust his package. I could have assured him it remained intact. Yowza.

Our little tete-a-tete is interrupted by the sight of Megan, the paramedic, walking toward us. Free of beekeeping protection today, I notice how attractive she is. Sean smiles at her. Megan's expression remains cordial until she looks at me and frowns. Megan is a member of the "lock her up" camp.

"His vitals are stable, but he is still confused. He's probably been lying out there for most of the night. Along with a head injury, he's dehydrated, probably contributing to his confusion. We've started an IV for fluids and plan to transfer him to the Lawrence Memorial for further evaluation. Is there any family we need to notify? Also, do you know his medical history or any medications he might be on? Does he have any allergies?" Megan asks. "And we'll need an emergency contact number."

"No," I reply. "Karl doesn't have any family around here. His wife died five years ago, and they never had children. I don't recall him mentioning allergies; he's always been healthy. I don't think he takes any meds. He uses the VA hospital in Topeka, so they might know. As far as emergency contacts, I guess I would be the closest. Let me give you my number. Can I ride with him to the hospital?"

"No," Both Sean and Megan said.

"There is no room," Megan says, her tone firm. "You'll have to make your own way, and don't try to follow us," she warns. "You could be ticketed for speeding."

"Really?" I ask. Boy, she dislikes me.

"Megan's right. Besides, I'll need you to walk through the—I believe you called it the honey room?" Sean says. "I want you to take a closer look. Let me know if anything is amiss. I'll drive you to the hospital myself after we're done."

I look between the two professionals. Megan's frown has deepened. "Why—what are you thinking? That this was an attempted burglary or something?" I ask.

"Right now, I'm not thinking anything. I need you to look around and see if anything is missing or looks out of place," Sean says.

"Okay—" I glance back at Megan. "But first, let me give you my number."

"Where are they taking him?"

Wyatt's question startles me as together we watch the ambulance doors close with Karl packaged up inside. "LMH. It's the closest emergency room that handles trauma. Normally, Karl uses the VA medical system, but we can sort it out later. I hope he is going to be okay."

"Me too," Wyatt says. "I hope you don't mind, but I went into the kitchen to find Elvis's dog food and water bowl. The poor ole guy acted like he was starving. He gobbled it down and drank an entire bowl of water."

I examine Wyatt's face closely and recall Sean's earlier question. Perhaps Sean was right. Just how well do I know him?

"Deputy Hernandez was with me, and I didn't touch anything other than the dog bowls and the dry kibble," Wyatt adds.

"How did you know where to find it?" I suddenly recall Wyatt's odd familiarity with the place upon our arrival.

"What? The kibble? It was in the pantry," Wyatt says. "I just—assumed. I mean—it seemed logical."

"Yeah—" Suddenly, I feel a chill of misgiving run through me. Could Wyatt? No, the notion was ridiculous. Wasn't it? Wyatt—a killer? Don't even go there, Indie. Instead, I focus on Sean. He raps his knuckles sharply on the back of the ambulance doors. The emergency lights begin to flash, and the vehicle rolls forward. "Sean—Interim Sheriff Riordan—needs me to take another look in the honey room." Odd. Why do I feel the need to correct myself in front of Wyatt?

"Do you want me to come with you?" Wyatt asks.

"Ah, no. Why don't you take Elvis back to The Hive. We'll need to look after him while Karl's in the hospital."

The sudden change in temperature from outside is noticeable, and I rub my bare arms. Sean was right. The windowless, utilitarian space resembles a storm shelter with thick concrete walls. A huge overhead ventilation system was designed to pull humidity out of the room and speed up the honey-drying process. It crystallizes readily when humidity exceeds eighteen percent. The oldest honey found dates back some five thousand years.

"Are you cold?" Sean asks, breaking the silence.

"No, I was just thinking about Karl. When we found him earlier, I feared he might be... dead. Who knows how long he'd been lying there? I don't like to think about what could have happened if we hadn't come along. Do you think someone did this to him?"

"That's a pretty plausible scenario since the door was locked behind him. We found a flashlight. It had rolled under the big stainless-steel tank over there. Mr. Bauman probably had it with him when he came out here. Unfortunately, the battery was dead."

"Karl must have come out here in the middle of the night to check on something—maybe something he heard? Or saw? And then someone struck him from behind, just like Professor Fontenoy. Thankfully, he's alive."

"That's a viable theory. Then again, it may have nothing to do with the Fontenoy homicide. He could have simply interrupted a burglary. Some of this equipment looks expensive, albeit cumbersome to steal."

"That's not too hard to steal." I point to the metal shelves holding a line of five-gallon buckets, each filled with fresh, lightly filtered honey. A noticeable gap is visible in the center of the second shelf. A space large enough to fit one of the containers. The buckets on either side appear to have shifted toward the center. Did someone break in here to steal honey? If so, it was a pretty pathetic attempt. "Who would break in here to steal a bucket of honey? Also, wouldn't you take more than one bucket?"

"Hmm... curious. Perhaps the perpetrator was trying to find something?" Sean walks toward the shelves. "Six buckets on the shelf above and only five on the one below." He slides one of the buckets to the side, widening the gap.

"There is something stuck on the back of the shelf, but my fingers are too big to reach it," Sean says. "See if you can grab the edge of it."

I peer into the space and notice a strip of dark green material caught on the back edge of the metal rack. I thread my fingers into the gap. "I've got it, but it's caught on the bucket handle below. Can you move it?"

Sean slides the bucket from the lower shelf onto the floor. I feel the strap tighten and then give way. I pull a length of green cord from the space. "It's a lanyard, but nothing is attached."

"Seems a bit coincidental, doesn't it? And likely not pertinent to our investigation. Who knows how long it's been stuck in there." Sean comments.

"Maybe not." I feel excited. Finally, a clue. "It's dark green, and the clip is shaped like a pair of praying hands, the color and emblem of the Colette University Saints."

"Hey, Riordan," Harlan Pierces yells. "You about done in there? We got a call I think you'll want to hear about."

Sean slips the lanyard into a clear plastic evidence bag. "Never a dull moment in this supposedly quiet county of yours." Exasperated, he shakes his head. "I'll be right out, Pierce."

Sean locks the door of the honey room and holds out his hand. Geez—it's like he doesn't trust me or something. Reluctantly, I pass over the key and follow Sean into the bright sunshine.

"What's up, Pierce?" Sean asks.

"We got a situation off Eisner Road," Harlan says.

"Eisner Road?" I speak up before Harlan can elaborate. Both men look annoyed. "I'm only asking because Eisner's property line backs up to my dad's."

"What's the situation?" Sean asks.

"Dispatch called and reported a group of drunk raccoons down by Eisner's cow pond," Harlan says. "They were stumbling around, acting crazy, and some babies died. It could be rabies. We need to take care of it, pronto!"

"*Procyonidae*," I murmur, not realizing I'm speaking the words aloud. "Known to carry rabies."

"What?" Harlan asks.

This time, the two men look at me questioningly. "I was saying raccoons are in the family *Procyonidae*. They are known as the washing bear for their propensity to wash their paws before ingesting food. So you said the gaze was already dead?"

"The what?" Harlan asks again.

"The gaze—the kits—the babies?"

112

"Yeah, whatever," Harlan sneers. "This isn't some episode of Kratts' Creatures, so why don't you speak English, DB? And why are you still here anyway?"

"Okay, okay, that's enough, Pierce. First off, did you notify Fish and Wildlife Services?" Sean asks.

"Yeah, I did, but the county only has one agent, and he's out on the county's side of Clinton Lake checking fishing licenses. It will be an hour before he can get the boat out of the water and head over here. Those coons could disappear into the woods, and we'll never be able to find them. My shotgun is locked and loaded. I say we drive out there and take care of the situation ourselves."

"I'm loath to agree, but Harlan could be right. Forty percent of wildlife cases of rabies are found in raccoons, so it's a strong possibility. But only a necropsy will tell us for sure."

"See, even DB agrees with me," Harlan says.

Sean speaks before I can open my mouth to call Harlan a jerk.

"Indie, since you are the closest thing we have to a biology expert right now, you'll ride with me to Eisner's pond. Then I'll drive you into Lawrence to check on Mr. Bauman. Maybe by then, Karl will be able to tell us something. Lead the way, Chief Deputy Pierce."

Harlan imitates the motion of racking a slide on a rifle. "Locked and loaded," he repeats.

Oh, brother. Once again, I follow the yin-yang law enforcement twins back to their respective vehicles.

"What do you think your sister will say when Wyatt shows up with that hound?" Sean asks.

Finally allowed to ride shotgun in the SUV, I'm busily fanning myself in front of the air-conditioning vent. "If I know Heather, she'll feed the old dog treats until he throws up. We've always wanted a dog."

"I'm surprised you don't have one. From what I can tell, you've got quite the menagerie."

"And that would be the problem. You've met our mother. She has an unhealthy affinity for exotic, furry creatures. Somehow, we always end up taking care of them. I don't think she cares for dogs. I reckon their coats aren't conducive to her weaving projects."

Sean grimaces. "I'd say your mother is certainly a force to be reckoned with, but maybe she was trying to share her love of natural fibers with her daughters. When you think about it, how much fiber can one long-haired rabbit possibly produce?"

"You would be surprised, and Mustard—" The words trail off as memories of the yellow and white striped kitten Claire had brought home resurface. Impossibly tiny and sweet, he'd been a rescue. As was Lettuce—the giant Angora who'd earned a reprieve from life as a breeding stud confined to a small wire cage. Satin and Lace were abandoned at a now-defunct llama farm in western Colorado. Heather and I love them all. "I guess I never thought of it from her perspective." In hindsight, I needed no excuse to blame my mother.

"My mom's a big gardener, and she would send me back to college with a bag full of groceries and some house plants. Somehow, I always managed to kill them," Sean says. "At least your mom's gifts have thrived."

"I guess so. Slow down—-Eisner's Road is coming up. The cow pond is back in those trees. I'm not sure Harlan can navigate his cruiser back there. It's a rugged cattle trail and still muddy after the rain we had the other night. At least, there are no cattle to contend with."

"Oh yeah, why not?"

"Tom Eisner died last year. Sally, his widow, sold off all the cows as none of their kids lived here. It was too hard to find labor. Unfortunately, it's the sad tale of many a small family farm. And even if you are willing and able to do the work, you're still at the mercy of Mother Nature. But then there is Big Ag, pricing you out of business. It's often a no-win situation."

"Yeah, I guess it would be hard to farm via a Zoom call," Sean quips. He slows the SUV and follows Harlan onto a deeply rutted dirt path overgrown with native grasses. "You weren't kidding about it being rough back here, were you?"

Looking ahead to the path as it narrows, I notice Harlan abruptly put on his brakes and stop the cruiser. He exits his vehicle and waves back toward us, shotgun in hand.

"We'll have to walk in from here," Harlan shouts.

"Harlan might be enjoying this a little too much." Sean sighs and stares through the windshield. "Lots of thick underbrush back here. No wonder I never saw a pond before."

"It's not a pond per se. I would call it more of a runoff ditch, a watering hole for cattle. Many small farms dig them in low-lying areas to collect drainage from creek beds and spring rains. Unfortunately, they're not a consistent water source, and by the end of summer, this one will dry up. But I'm glad I brought along my faithful muck boots." I swap out my white canvas sneakers and leave them tucked neatly into the floor well.

"Yeah, lucky you," Sean mutters. He glares down at his polished brown cowboy boots. "Guess I know what I'll be doing tonight." Then he pushes his Stetson down firmly on his head and opens the door. "Let's get this over with."

Besides the rising temperature, I first notice the smell as I step out of the SUV. There is a slight breeze, but from this distance, a pungent chemical odor and the stench of rotting vegetation permeate the air. I notice Sean pull a tan handkerchief from his pocket and place it over his nose and mouth.

"Good Lord, what is that smell? Something must be dead back there," Sean says.

Harlan slips a black balaclava over the lower half of his face ahead of us. He proceeds down the path in a crouch, his rifle tip swinging from side to side. I agree with Sean; Harlan is enjoying himself way too much. "The underlying scent is chemical, combined with decaying organic material. I bet you dollars to donuts those raccoons aren't rabid at all. They've been poisoned."

In the end, it wasn't the smell that got me. Oh, it was awful, but not nearly as bad as seeing two adult raccoons and three kits lying dead by the edge of the watering hole.

Sean sports a pensive expression, and Harlan remains uncharacteristically quiet as we survey the carnage. Finally, Sean asks. "Am I to assume this is an illegal dump site?"

"Not necessarily." I point to the sloping ground behind us. "Remember I told you these cow ponds are reservoirs for runoff, so who knows where the tainted water originated from? We need to check a topographical map for the nearest elevated water source. There may well be other sites of contamination."

"Indie's right," Harlan says. "We should talk with Mrs. Eisner, and then I'll notify the Fish and Wildlife Department. They are in charge of pond management in the county. This could be bad. I imagine the feds will eventually get involved."

"The EPA?" Surprised, I look at Harlan with newfound respect. He was spot-on in his assessment.

"Well, I know one thing—this is way above my pay grade," Sean says. "Chief Deputy Pierce, since you are familiar with the property owners, I'm putting you in charge of this case, so proceed how you think best."

"Yes sir," Harlan snaps a salute.

I watch Sean slap Harlan on the back. Reluctant to interrupt this rare moment of male camaraderie, I hesitate. "Ah, guys—just one more thing."

"What?" Sean asks.

"Yeah, what now, DB?" Harlan challenges.

In the spirit of our newfound cooperation and my desire to get out of there, I let the misnomer slide. "I'm pretty sure I recognize the odor." I pinch the bridge of my nose to deaden the effects of the smell. Both men stare at me like I've grown a head full of snakes. But even Medusa's mythical serpents would be dead from exposure to this toxic slew. "Along with the obvious smell of decomposition, I'm pretty sure the underlying smell is a concentrated form of ATI/EX 50/50. So, Harlan, you might want to swing by Agro-Tech, and talk to Josh Blake about their waste disposal procedures."

"Roger that, DB," Harlan says.

Chapter Fifteen

To say the drive to Lawrence Memorial Hospital was anticlimactic is an understatement. Sean drives swiftly, once again exceeding the county speed limits. Besides the police radio chatter, we travel in silence.

For the first few miles, Sean opens the windows to blow out the stink of chemical decomposition. But now, the air conditioning is on full blast in the quiet interior. I notice, for the first time, that he is not wearing his black eyepatch. The gleaming blue orb of his artificial eye slides over me. He must see my surprise and touches his brow subconsciously.

"You got a theory on what we just found out there?" Sean asks.

Wearily, I allow my head to rest against the side window. "Yes, unfortunately, I do."

"Pardon my pun, but are you gonna spill it or what?" Sean asks. "I've never known you to be reticent about voicing your opinion."

"I'm just reluctant to rat out a friend—especially when I don't know all the extenuating circumstances."

"Go on. I'm listening."

"I don't think you will like this, but I visited Agro-Tech Industries yesterday. Josh Blake, the shift foreman, is an old high-school friend. I might have convinced him to give me a quick tour of the facility. That's how I recognized the smell. It took me three times to shampoo the stench out of my hair."

Sean lets out an exasperated sigh. "Let me guess, and you were looking for Carson Wells?"

"Yes, how did you know?"

"How many times do I need to tell you, Indie, I know what I'm doing. Of course, the Sheriff's Office is investigating the connection between Agro-Tech and the grant awarded to Colette University. But there is a right and wrong way to conduct an investigation. And the correct way doesn't include a civilian barging into a place of business asking questions they have no right to ask."

"I know, I know, it was stupid of me. As it turns out, Carson Wells wasn't even there. He was picking up the Agro-Tech executives before last night's disastrous Chamber meeting. Can you imagine how awkward the drive back to the airport must have been? Besides, Josh didn't think twice about me asking questions. I told you—we're old friends."

"Oh yeah, and you don't think Josh Blake would mention to his boyfriend that an old high school chum dropped by out of the blue and started asking a bunch of nosy questions about the plant? And as it just so happens, a said friend recently found the chair of the university's Science Department dead from an apparent homicide. No, there is nothing suspicious about those circumstances," Sean says.

"Wha—whoa—wait a minute! Did you say Josh Blake's boyfriend? Are you saying Carson Wells and Josh Blake are—?"

"Lovers? Yes, Indie, that is exactly what I'm saying. For Pete's sake, you must get your head out of the sand and stop interfering in things that are not your business. Pay attention to what's going on around you."

Ouch. That was harsh. But how had I not known? Josh was—is my friend. Had I always been this obtuse and superficial with my friendships? My recollections were of the wise-cracking kid who'd lived in a home with an absent mother and a father known for his volatile temper. It could not have been easy coming out as a gay man in our small, rural community. And realizing I'd not been the kind of friend Josh could share his true feelings with made me feel considerably worse.

"Hey, Indie, snap out of it. We're here at the hospital. Hopefully, Karl will be able to tell us who attacked him, but you need to let me ask the questions. You got that?"

We find Karl under observation in the emergency room. According to his nurse, his initial CT scan was negative for any sign of bleeding. Still, due to his age and the questionable circumstances surrounding his injury, he is awaiting a transfer to ICU.

The ICU waiting room is practically empty when I take a seat. Sean and I watch the older man roll past us on a gurney. Engulfed in a tangle of wires and tubes running from various monitors, Karl appears shaken and pale against the white sheets. The staff pushes through the automated doors behind the *Authorized Personal Only* sign without pausing.

True to my word, I said very little in the ER, deferring to Sean. It was difficult to keep quiet, but I felt terrible seeing the older man in such bad shape. The nursing staff cautioned me it would be a while before I could see him and only for brief intervals. Ostensibly searching for coffee, Sean moves out of earshot when his phone rings.

Once again, I am swamped by guilt. Who had done such a thing? Was I somehow to blame? And how exactly did a Colette lanyard end up wedged behind a honey shelf in Karl's bee room? Did the green lanyard belong to the perpetrator? Or was it planted evidence, signs of a more sinister plot? Perhaps, an effort to implicate Karl or someone else? So many questions and, thus far, so few answers.

I look up to see Wyatt walk into the waiting room, followed by none other than Granny Bea.

"Wyatt! What are you doing here? Is everything okay?"

"Yeah—I mean, yes, everything is fine. I dropped Elvis off with Heather. When I left, he was taste-testing a new doggie cookie Angela baked. I think he is enjoying all the attention. I thought you might need your car, so I drove it over. Granny Bea had errands to run in Lawrence. She'll give me a ride back to Colette. How is Mr. Bauman?" Wyatt asks.

"Oh, that was nice of you both." It was. Completely unnecessary but nice nonetheless. "Karl is conscious. His vitals are stable. Unfortunately, he'll need to remain in the ICU overnight. Concussions can be dangerous at his age. I've only been able to see him briefly. The visiting times are limited to one person and only for a few minutes."

"I can wait with you," Wyatt insists.

"Me too," Bea says.

"Oh, there is no need for that. It's fine. Sean, I mean, Sheriff Riordan is around here somewhere."

"Price, what are you doing here?" Sean asks, his voice filling the space.

Sean hands me one of the steaming Styrofoam cups of coffee and then inserts himself into the space between us. The gesture is deliberately protective. Wyatt takes an involuntary step back.

"Ah... there he is—the Knight of Swords," Bea murmurs.

"What?" Sean turns to Bea. "Oh, I didn't see you there, Granny Bea. I hope it's not inconvenient, but Indie and I need a private word with Mr. Price."

Do we?

A nurse in lavender scrubs steps out from behind the ICU's automatic doors. "Family of Karl Bauman?"

I start forward, but Sean places a restraining hand on my arm. "Indie, I would prefer you to stay here and be a part of this conversation," he says in a low voice.

"But I—"

"I'm family," Bea pipes up.

I start to object, but Sean tightens his hold.

"Are you his wife?" the nurse asks, her tone full of empathy.

"I'm his fiancé, dear," Bea says brightly. "How is he doing?"

Bea makes to follow the nurse through the automated doors. She stops to whisper to me, "Remember nothing is as it seems." Then, turning to Wyatt, she says, "*Veritas vos liberabit.*"

"Let Bea go in," Sean says. "I need you here."

"Sean," I protest.

The look on Sean's face cuts off any further challenge on my part. Helpless, I watch Bea disappear through the automated door.

"So, Wyatt T. Price of Fort Wayne, Indiana do you want to explain why you decided to apply for law school in Topeka, Kansas? So far from home, after finishing your undergrad at Indiana State? I believe your mother still resides in Fort Wayne?" Sean asks. "Her name is Sarah Price? And your father was Wayne Price, now deceased having died in a drunk-driving incident when you were an infant? You survived the crash, but the other car's driver, a mother of two young children, did not."

I look at Wyatt with growing horror. Not that he'd had much, to begin with, but now the young man's face is completely drained of color. Crash? Wyatt hadn't told me his father had been driving drunk and that he had been in the car. What the heck was going on here?

"I can explain," Wyatt says hurriedly.

"I would hope so," Sean says. "And while you are at it, maybe you can explain why you seem nearby every time a violent crime occurs?"

"What—? We—wait—you think I'm involved in the murder of Professor Fontenoy and hitting Karl Bauman in the head? I would never—" Incredulity fills Wyatt's voice.

"I don't know, Wyatt, you tell me," Sean says. "It is suspicious how you always seem to be around. Somebody had to commit these crimes? What do you have to say about that?"

"No, I… I swear to you, Sheriff. I had nothing to do with any of it!" Wyatt says. "I— would never hurt anyone."

Sean does not respond. For once, I'm stunned into silence.

Wyatt, eyes wide, turns to look at me. "Indie—you believe me, don't you?"

"Wyatt… I don't know what to say. Why didn't you tell me your father was the drunk driver? You said he was a good guy."

"I shouldn't have told you that," Wyatt concedes. "You heard the sheriff—I was just a baby! I suppose I wanted to believe he wasn't a terrible person. It's not exactly something you bring up in polite conversation. 'Hi, my name is Wyatt, and my drunk father killed some other kid's mom while I survived strapped in my car seat,'" Wyatt exclaims. "But I swear to you, Indie, I had nothing to do with killing Professor Fontenoy, and I would never hurt Mr. Bauman. I—I was trying to help. Please say you believe me. It's not what it seems."

Nothing is as it seems. Bea had whispered those exact words. What was that about? On the other hand, being accused of a crime, I didn't commit gives one a whole new perspective. So, I made a decision. "I believe you, Wyatt."

"You do?" Wyatt asks.

"I do."

"Hey," Sean says, "you're not the one he needs to convince. Or did you forget? I'm the investigator here."

"I remember. But, Sean, you believed me when no one else did. Now I choose to trust Wyatt." I stand next to Wyatt to face Sean in a wall of solidarity. "Just tell us the truth, Wyatt."

"Good news," Bea exclaims, bustling back through the ICU doors. "My fiancé is awake! Also, Wyatt, I'm positive he is your grandfather."

In unison, the three of us gaze at the tiny woman.

Sean shakes his head. "No one ever told me the people in this county were straight-up delusional."

"You think so, Bea?" Wyatt asks. The color in his face returns, his words tinged with excitement.

"I do," Bea says. "He has the same crooked pinkie finger as you do."

"Whoa—whoa, wait a minute! What's this about Karl being Wyatt's grandfather? And, Bea, you know you are not his fiancé, right?"

"Of course I do. But now I'm on the approved visitor list, so I can be here if you can't," Bea says. "Besides, I pulled the Page of Cups this morning. Love is in the air. Karl Bauman is a remarkably fit man for his age, don't you think?"

How did this conversation manage to fall off the rails so quickly?

"Are you saying you're related to Karl Bauman?" Sean asks, focusing on Wyatt. "How so?"

Wyatt's complexion flushes to bright red. He clears his throat and looks at Bea. She smiles at him, nodding encouragingly.

"Just tell them what you told me, dear," Bea says.

"Indie, do you remember I told you my grandmother passed away a few years ago?"

"Yes…."

"I was helping my mom clean out my grandmother's old papers after she passed. We came across a bundle of letters. They were unopened and labeled *Return to Sender*." The letters were all addressed to a PFC, Karl Bauman, Army Air Force, Postal Service, APO 1143, Saigon. Naturally, I was intrigued, having never heard the name Karl Bauman before. Unfortunately, those air mailers are tissue-thin, the writing old and faded, so it took me a while to steam them open.

"Interesting, but what did the letters to Karl say?" I interrupt, still not making the connection.

"We knew my grandmother was born in Kansas, but she never spoke much about her family here. I guess my grandfather, or at least the man I thought was my grandfather, Thomas Price was from Kansas City, Kansas. He was fifteen years older than my grandmother and a diabetic, so he missed out on the draft. My grandmother always told us they married after a whirlwind romance and then moved to Indiana for a job. As it turns out, my grandmother was pregnant but not by Tom Price.

"Go on." Now, I'm intrigued.

"The letters revealed my grandmother was pregnant by Karl. I told you the letters were returned unopened, so I doubt Karl ever received them. It was the fall of 1967, during the buildup of the Tet Offensive. The U.S. Army moved troops around frequently, and the military mail had difficulty catching up with PFC Karl Bauman. As the weeks turned into months without a word from Karl, her missives became increasingly anxious. I think she feared the worst when the packet of letters returned unopened,"

"Did she think Karl had abandoned her? You seriously can't believe that!"

"No, I think she thought he was dead. We are talking about 1967, and the news broadcasted daily casualty counts. At nineteen, my grandmother was pregnant, unwed, and from a traditional religious family. Look, what I told you about my grandfather was all true. He was a standup guy. As far as he was concerned, my mom was his daughter, and I was his only grandson. We moved in with my grandparents after my father died. So, you can imagine, my grandmother's revelation shocked us both, and my mom begged me to leave it alone. She didn't want to dishonor the memory of her adoptive father.

"But you didn't leave it alone… did you?" Bea asks.

"No." Wyatt shakes his head. "I did not—and my mom still thinks this is all a wild goose chase."

Sean exhibits a faint air of skepticism while taking in Wyatt's story. Thankfully, he remains silent, which allows me to continue questioning Wyatt uninterrupted. "Okay, hypothetically, let's just say this story is true. It still doesn't explain how you ended up here in Colette and why you have yet to tell Karl this story."

"Do you realize how little presence a man of Karl's generation has on the internet—— no social media at all? Honestly, I thought he'd died in Vietnam. Why else would those letters be returned unopened?"

"Did you check the National Achieves?" Sean asks.

"I did," Wyatt says. "Did you know there were over 50,000 American casualties during the Vietnam War? It took some time to comb through the Vietnam Veterans Association archives. I never found Karl Bauman on the missing or killed list, so I had to start with the premise that he survived. I figured he would return home to Kansas after the war," Wyatt says.

"Logical," Sean says, speaking for the first time.

The wary look on Wyatt's face changes to something akin to gratitude.

"But how did you get here to Colette?" As interesting as all this information is, I need Wyatt to get to the point of this convoluted tale.

"On impulse, I applied to Washburn Law School in Topeka. Being in the capital city, I assumed I'd have greater access to historical records. My mom was not too happy about my decision to move to Kansas, but Washburn is a great school, and I felt I owed it to my grandmother to continue the search. She'd sacrificed a lot, even saving for my college education. Plus, my grandma kept those letters for a reason. I felt she would want me to try at least to find Karl. I narrowed my focus to Corley County when I found an old college certificate of hers from 1959. After graduating from a two-year business course, she worked at the First Bank of Colette. Coincidentally, my grandfather was the assistant bank manager at the time. That is how they first met."

"Really? My grandmother worked at that bank. Maybe they knew each other?"

"Maybe, but I don't think she worked there too long. She and Karl Bauman were secretly dating before being deployed to Vietnam. I say secret because my grandmother's family were staunch Catholics, and the Bauman family was German Lutheran, so her family disapproved."

"I'm sure they didn't approve of having a baby out of wedlock either," Bea pipes up.

"For sure, the reason why my grandmother's wedding to Thomas Price was so rushed. As I said, he was quite a bit older than her and Catholic. And before you think she pulled the wool over his eyes, he knew about the pregnancy and married her anyway. They grew to love each other, and Grandma was devastated when he passed away. We all were," Wyatt says. "When I started volunteering at your dad's student law clinic, I heard him mention he was from Colette—"

"Are you telling me my dad knew about this?"

"No—I told him I was tracing my family roots, but I didn't feel comfortable mentioning the Karl Bauman connection yet. I decided to visit Colette and check out the area. I wandered into The Hive for coffee and to use the Wi-Fi. I met Heather because she had some issues with the Wi-Fi router, and I offered to help. But I swear I didn't know she and you were Professor Evans' daughters. Once we got it sorted out, Heather offered me a job. The rest is history."

"Yeah, right—the age-old scenario of a boy wandering into a shop and becoming enamored with the shopkeeper. And now he doesn't know how to tell her he's been there on false pretenses." No wonder the guy is always acting so ill at ease. I could slap him.

"I wanted to say something when I learned Karl was your beekeeping mentor, but I knew you'd been having a rough time with your recent—"

"Breakup," Bea finishes.

"Ah—yeah—it never seemed like the right time to say anything. Until today, that is, when we found Karl unconscious…" Wyatt shudders. "My first thought was that I might have lost my grandfather before I ever had a chance to meet him."

"The truth shall set you free," Bea says. *Veritas vos liberabit.*

I swear that woman is always two steps ahead of the rest of us. There may be something to all this woo-woo stuff after all.

"You could still be the one who killed Professor Fontenoy," Sean challenges. "Just where were you on the morning of his murder?"

"I swear I have an alibi," Wyatt asserts. "Professor Evans had me drop off some paperwork with Miguel and Rosa Gutierrez. As you know, the law clinic has taken up the asylum plea. Rosa cooked a huge breakfast that morning and insisted I stay and eat. She says I'm too skinny." Wyatt flushes. "I was there promptly at 7:30 and back at The Hive by 9:00 to meet Bea to set up her social media pages. It takes me forty-five minutes to drive from Topeka. My roommates can verify the time I left."

"He was there at 9 o'clock sharp," Bea says. "I always say being on time is an attractive quality in a man."

I examine Wyatt closely. Rosa is correct: he is too skinny. He is tall and thin, his features too big for his face, not unlike Karl's. And, as Bea correctly pointed out, he suffers from the same minor bone malformation, clinodactyly, on his right pinkie finger as Karl does.

"Hmph," Sean says. "I'll speak to Rosa."

"Bauman family?" The same lavender-clad nurse appears. "Mr. Bauman is asking to speak to... Indigo Bleu? Am I pronouncing that correctly?"

"Yes, that's me." My head is beginning to pound, but I feel fiercely protective of the older man. Though it's probably from drinking coffee on an empty stomach, I suspect Wyatt's revelations have again sent my emotions into a tailspin. I give Wyatt and Bea a stern look. "Don't you dare tell Karl. He needs to recover before he hears this story, and I want to be there when you do."

"I need to get back to Colette," Sean says abruptly. "But Indie's right. Not a word. The man has a head injury."

"Agreed," Wyatt says.

Bea makes the universal pantomime of zipping her lips and turning a key. The skeptic in me and the twinkle in her eyes have me doubting her intentions. Bea is dressed in all blue today, and I'm afraid to ask what the color implies.

"I want to meet my grandson," Karl says. His blue eyes flare. Lifting his head from his pillow, he yanks the oxygen tubing from his nose.

"Karl! What are you doing?" The man is as stubborn as a mule. Even a concussion can't keep the crusty old beekeeper down. And, as it turns out, a blue-clad sprite of a woman cannot keep a secret. "You need to rest."

"I'll have plenty of time to rest when I'm dead," Karl replies. "If I have a grandson out there, I want to meet him. Besides, why are you still here? Shouldn't you be out trying to figure out who did this to me? It's bound to be the same person who killed your professor. You send that woman in here who claims to be my fiancé. She's got enough wind in her sails to keep the ship afloat around here while you're gone."

Chapter Sixteen

B ea and I leave Wyatt curled in one of the uncomfortable ICU waiting room chairs. He'd insisted on staying with Karl; no amount of arguing would change his mind. Finally, one of the nurses took pity on Wyatt and gave him a pillow and blanket. I'd followed Bea home from the hospital and slept without dreaming for the first night in several days. Heather had acquiesced, albeit grudgingly, to care for the old hound. In exchange, I had promised to work the shop this morning.

So, I'm surprised to hear a flurry of sharp barks followed by the distinctive wail of a basset hound as I make my way down the back stairs of our apartment. Uh-oh. This sudden outburst of activity cannot be good. I pass Angela's kiosk. It's deserted—time to pay the piper.

I walk into chaos. Two stacked bins of our signature goat milk soaps lie upended, the bars scattered across the floor. Broom in hand, Angela is sweeping up a pile of cookie crumbs. Who knows whether they are of canine or human variety, but at this point, does it matter? Elvis lets out another signature wail, followed by the yowl of an indignant cat perched above me. Great.

"Elvis," Heather cries. "Get down from there. It's okay, Mustard. He is just trying to be friends."

Friends? Yeah, right. I spy Elvis, his nose pointed straight toward the ceiling, and his front paws and low body stretched to elongated proportions against the shelves of honey. One solitary plastic bottle plops from the top shelf as if in slow motion, almost hitting Elvis square on his nose. It's followed closely by another. This one hits the floor nearby with a thud. I see Mustard strategically pawing the golden projectiles onto the hapless dog one by one.

"Indie, grab the dog. He has poor Mustard treed," Heather yells.

"Girls, what in the world is going on in here? Indie, you need to do something."

"*Mamma mia!*"

Of course, my mother and her Italian sculptor friend decided to visit the shop this morning. Giovanni is not wrong about calling the scene one of lunacy. But for my mother to assume I should rectify the situation is beyond the pale. It's also standard operating procedure for Claire.

"Ow! Get him off me." A muffled yelp escapes Claire's lips as Elvis transfers his affections and slobber onto the front of her floor-length boho skirt and to the ends of the delicate open-weave scarf draped around her neck. "Indie, this scarf is angora and handwoven by an artisan in Anatolia. Ethically sourced, of course."

"Well, that explains it. It smells like a bunny rabbit. Speaking of rabbits, Heather, please tell me Lettuce is his cage." I walk over to collar the snuffling hound and drag him away from my mother. I can't help but take satisfaction at the string of doggie drool plastered on the hem of her skirt.

"Oh, Elvis and Lettuce are firm friends already. That's why I thought he would be okay with Mustard. I was mistaken," Heather says. "I'm not sure Elvis can stay here. Mustard is pretty upset."

Duh, you think? "Well, Mustard is going to have to adapt. At least until Karl gets out of the hospital. Unless…" I look meaningfully toward my mother.

"Good heavens, no!" Claire exclaims. "I can't have that slobbering beast in my home."

"*Mio Dio*, no!" Giovanni echoes the sentiment.

I narrow my eyes to examine the pair more closely. Is Giovanni living in my mother's house? And if so, does that mean—? Ugh. Contemplating the implication was too much. "I'll take Elvis upstairs. But don't leave, Claire. I have a few questions to ask you."

After inhaling a bowl of kibble, Elvis snoozes peacefully on an old blanket in the kitchen. Wreaking havoc in the shop is exhausting work.

Retracing my steps down the backstairs, I spy Claire and Giovanni seated at a table in front of Angela's kiosk. Two tiny espresso cups sit on the table in front of them. I should have drawn their attention to my return, but instead, I secret myself at the base of the stairwell. With equal parts aversion and fascination, I notice Claire flush as Giovanni reaches across the table and runs his strong fingers across the top of my mother's hand. Then my mouth falls open as the sculptor leans in to kiss my mother passionately on her lips.

I stay hidden until Giovanni pushes back his chair and strolls out the front entrance of The Hive. Claire stares longingly at his retreating backside. Oy! I have to admit it is a mighty fine sight.

But, as long as I live, I will never unsee the moment's intimacy. Yes, the divinely handsome sculptor is living in my mother's house. What must Heather be thinking, and speaking of Heather, where has she gotten off to?

"Mother, where is Heather?"

Startled, Claire looks up, a bemused expression across her face. The contents of the demitasse cup in front of her remain untouched, which is good. On a good day, I can't handle my mother, least of all, jacked up on caffeine. But after the familiarity I'd witnessed between the pair, I could undoubtedly use a jolt.

"Are you going to drink that?"

"What?" Claire looks down and shudders. "No, it's too strong."

Picking up the delicate cup, I down the contents in one gulp. My mother's eyes widen in surprise. I repeat my earlier question: "Mother, where is Heather?"

"Oh, she left to run errands," Claire says, her tone vague. "Heather said you promised to mind the store this morning. I told her I would wait until you got that—drooling hound settled." Claire shudders again. "Whatever compelled you to bring a dog in here?"

"Elvis belongs to Karl Bauman. He is still in the hospital after being attacked yesterday."

"Oh, I heard about poor Karl—such a dear man. On the bright side, Heather says he will make a full recovery. And since Giovanni is spending the day in his studio, I'm here to open Oaxaca as Mari has the day off."

"Yeah, thankfully, Karl will make a full recovery." Wow, empathy was strong with this one. Not. I need to get in front of this conversation quickly.

"Mother?"

"Yes?" Claire responds, rising from her chair.

"What's going on between you and Signor Vitale?"

"I'm not sure what you mean by that question, Indigo?"

"Mother, don't try and deny it. I saw Giovanni kiss you goodbye, and it wasn't your traditional farewell, Italian or not. So, I will ask you again—what is happening between you and the very charming but much younger Giovanni Vitale?"

"European men don't get hung up on age, and not that it's any of your business, but it's only a ten-year difference. Our patriarchal society puts these absurd conditions on what we deem appropriate. For example, let's reverse the situations—if your father were dating a younger woman, you wouldn't be asking these demeaning questions," Claire expostulates.

For once, she was correct. "Wait—you don't think Dad is seeing someone younger, do you?"

"I wouldn't know. If you recall, your father and I divorced last year," Claire says coolly. "But I wouldn't be surprised."

"You wouldn't?"

Claire must have seen the expression on my face. Her tone softens immediately.

"Indie, I know the situation with your father has been especially hard on you. Neither one of us has been exactly forthcoming with you girls. But I refuse to say anything to jeopardize your relationship with Daniel and think it best you discuss the matter with him when he returns. All you need to know for now is that your father and I wish only the best for our daughters. And that includes your safety and well-being, so I feel confident speaking for both your father and myself. This obsession with murder must cease."

Seriously? I stare back at my mother defiantly. She is the first to look away. "You know I can't do that, Mother. But I appreciate your concern. I'm here, and I've got the shop covered. So, why don't you do whatever you need to do."

"Indie—" Claire says, her eyes welling with tears.

"Please, Mom… don't say another word." Unexpectedly, I also feel tears in my eyes. Darn, that is twice this week. I turn away before my mother can see them.

Chapter Seventeen

The Hive remains all but deserted this morning. Despite the espresso shot, my conversation with Claire has drained me emotionally. Once outside, I notice a shiny black Range Rover pull into a parking space across the street. I recognize the vehicle as belonging to Carson Wells. The choice of the expensive car confirms my suspicion that Carson is not from here. I watch Carson Wells and Josh Blake pick their way across the street toward the front entrance of the Sheriff's Office.

"Josh!" I jog quickly down the sidewalk along Main. Josh glances my way and starts to wave. Carson Wells grabs Josh by the shoulder and pushes him through the heavy metal door. They both disappear inside. So much for my chance of getting the first crack at questioning them.

"I punch in Sean's number. He answers on the second ring, "Indie, I told you to stay out of this." Silence follows on the other end.

Sean Riordan did not just hang up on me. Did he? I look down at my phone in disbelief, longing for the days of satisfaction you received by slamming an old-fashioned handset back into its cradle. Instead, the sleek black screen gleams back at me, utterly devoid of sound. Sean Riordan just told me to stay out of it. Not bloody likely. I start to hit redial but think better of it.

"Screw it." Running down the sidewalk, I thrust open the Sheriff's Department's door and smile politely at the young deputy manning the metal detector

131

at the entrance. I throw my car keys and cell phone into the scanner, a necessary evil even the most minor police department has been forced to adopt.

Approaching the duty desk, I greet one of the county dispatchers, Miranda Ledbecker. Miranda, or Randi, as she prefers to be called, is a contemporary of my mother. But, unlike Claire, Miranda looks and mainly acts her age. "Hey, Randi, how have you been?"

Randi looks up from the clear plastic salad container on the front desk. She sighs and pushes it to the side, equal parts longing and distaste on her face.

"Fine, or I will be as soon as I can finally eat lunch, such as it is." Randi pats her thick middle. "Not everyone can stay as trim as your mother. How is she, by the way? I saw her the other day with some long-haired hunk. What's up with that?"

"Ah, that was Giovanni Vitale, the new artist in residence from Colette. He's a sculptor. Claire's been helping him settle in," I say, feeling slightly defensive.

"Oh yeah, Claire looked pretty settled if you ask me. Your mother always had good taste. Look at your dad. He's a total stud for his age. And speaking of studs, your name crossed the lips of our new interim sheriff lately, and not in a particularly good way. So, if you plan on speaking with him, you better keep your guard up. He is not in the best mood today. If I were you—"

Ugh, my dad is a stud? Uncomfortable with the direction of this conversation, I interrupt Randi, "Uh, well, I was hoping to speak with him anyway. Can I go back to his office?"

The phone on Randi's desk begins to ring, and she pushes aside the salad container and waves me down the hallway to the right.

I pass by a glass-fronted cubicle where Deputies Nolan and Hernandez sit behind desks facing each other. I had yet to see one without the other.

"Can I help you with something?" Hernandez calls out through the open doorway.

I stick my head inside. "Did I just see Carson Wells and Josh Blake pass through here?"

Hernandez looks over at Nolan, who doesn't respond. "Yes, they did. Is there something we can help you with?"

"No, not really. I just wanted to—" Down the hall, I notice Sean Riordan exit a door and close it behind him. Intent on his phone screen, he fails to see me and proceeds down the hall.

"Sean." I hurry past the two deputies before Sean can enter the adjacent room. Behind me, Nolan and Hernandez don't bother to get up. Upon hearing his name, Sean turns around, his hand poised on the door handle.

"Indigo. What are you doing here?" Irritated, he glances back toward the closed door he'd exited and then to the one in front of him. "Oh, let me guess—"

"Sean, I was hoping you would let me sit in while you question Josh Blake?"

"What? Absolutely not!" Sean says. "Have you forgotten you are neither law enforcement nor legal counsel? So, you are not entitled to barge in here and demand to be included in a person of interest interview."

"I know, but—wait—Josh is a person of interest? Then you have to let me speak with him. He's my friend and may know more than he thinks he does. I could be useful."

"Useful? Are you so sure about that? Or are you forgetting you're the one who outed your so-called friend who, by all accounts, made a mistake that wasn't just dumb but illegal when he ordered the contents of those barrels be dumped on private property," Sean says. "I'm not sure how happy he will be to see you again."

"True," I agree. "But suppose Carson Wells knew about the flawed data and used it anyway. Josh might be more likely to tell me."

"Indie, when are you going to get it through your head? A murder investigation is strictly about facts, not feelings or suppositions. Being friends with Josh will not make him cooperate if he tries to hide something," Sean says. "Now, I'm going to go in there to interview him alone, and you need to turn around and go home, Indie!"

Turning the knob, Sean steps through the doorway before I can ascertain who is inside. It closes behind him with a pronounced click. An air of censure emanates from the silent deputies. I perp-walk past them toward the exit. I wave half-heartedly to Randi, thankful Chief Deputy Harlan Pierce isn't around to taunt me.

Sean Riordan practically slammed the door in my face. The man was beyond infuriating. He was accommodating when it suited him, but now, he blocks me from the investigation. I needed to question Josh Blake about the runoff fiasco myself. After all, I have environmental concerns, too. Oh, who was I kidding? My sole objective was to glean any information pertinent to my most pressing question: who killed Professor Fontenoy? And I wanted to do so before Sean did.

So, Indie, what's your next move now? Maybe the killer wasn't Carson Wells, but I am convinced I am getting close to the truth. The whodunnit spotlight would have to turn to the trio of Colette academics and Brenda Matthews.

I spy Brenda Matthews getting into her silver Lexus SUV. Sean Riordan may have scoffed at the possibility of Brenda's involvement, but I'm sure not. Still fuming from Sean's verbal dressing down, I race back to slide into my Subaru parked in front of The Hive—Heather's missive to mind the store wholly forgotten.

I pull in behind Brenda. She drives down Main Street well over the speed limit. Where is a cop car when you need one? I don't have the time for this, so I make an illegal U-turn in the middle of the street. As soon as we pass the city limits, I am forced to speed up to keep up with her. The mayor's wife has a lead foot.

Once again, I notice a loofah dangling from her rearview mirror. It's black. Weird. I remember seeing one of those silly things hanging in her car before. Had it been pink or white.? Pink, I recall. What is up with that?

Brenda takes the road leading west out of town. She accelerates on the straight, open road, and I trail a short distance behind. Still, my speedometer edges past seventy. Where are you going, Brenda? Late to a property showing? Or, in a rush to commit another murder?

Since vehicular surveillance is not my bailiwick, I hang back and set the cruise control to a respectable sixty-seven mph. Traffic is scarce, and I keep our intrepid realtor in sight. At this rate, we'll reach the Douglas County line. She could be taking a circuitous route into Lawrence. If so, should I swing by the hospital to check on Karl? No, Karl had the full attention and care of his newly found grandson, Wyatt T. Price.

Of all the craziness of the past couple of days, Karl being related to Wyatt is a scenario I never envisioned. But Karl appears delighted, even ecstatic, over the possibility of gaining a grandson. It must have something to do with Bea. I would never admit it, but the tiny woman seems to have serious mojo. I was happy for them. Everyone deserves a chance at love and family.

The gas gauge indicates I have a quarter of a fuel tank. Surely Sean has finished questioning Carson Wells and Josh Blake by now? I should have stayed behind and waited for them to re-emerge. There was zero chance of getting any information from the tight-lipped Carson, but Josh might inadvertently blurt out something of interest. However, the possibility of doing so has faded fast with my current speed and distance from town.

At last, Brenda hits the brakes. Her car slows, and the left turn signal comes on. My curiosity peaks as I follow her onto Baker Road. The four-way stop sign for US-59 and the county line is coming up. It's hardly a surprise, but Brenda barely hesitates as she blows through the intersection.

Indeed, she must be going to Lawrence. I consider breaking off my pursuit. This trip is turning into a big, nothing burger. Once again, I see the blinker indicate another left-hand turn. The SUV swerves onto what appears to be a private gravel drive, and Brenda accelerates. I noticed a small placard planted on the right side of the shallow trench: *Lakeview Lots for Sale*. The sign has a large red arrow pointing straight ahead. A cluster of loofahs of various colors is tied onto the sign's post. White, purple, yellow, pink, and blue loofahs surround a sizable black scrubby. They flap about in the breeze like a grotesque caricature of a cheap bouquet. What is this, some weird bridal shower?

I'm committed, so I cautiously follow the dust from Brenda's car around a long, curving drive lined with newly planted trees. The road ends at a construction site. The house before me is palatial, a white- and black-trimmed, modern-style farmhouse. Even with partially completed landscaping, the place is stunning. Maybe she is here for a listing after all? Brenda parks in front of a three-car garage alongside a black Mercedes SUV. One blue and one yellow loofah hangs conspicuously from the rear bumper of the expensive vehicle. Okay, this is seriously starting to creep me out. I park some distance from the house. Another car roars past me, and a purple loofah flaps wildly from the driver's side window.

I hit Sean's number on speed dial, and it rings twice before I hear his voice.

"Indigo?" Sean asks. "I hope you're not calling to weasel any information from me. Because I'm not going to—"

"Sean—" I whisper into the phone. "Please listen to me. I'm calling—"

"You need to speak up. I can barely hear you. Why are you whispering?" Sean asks. "Where are you?"

Why am I whispering inside the car with the motor running and the air conditioning on full blast? "Sean, just listen to me. I'm dropping you my location pin. I followed Brenda Matthews. I don't know what's going on, but—"

"You what!" Sean shouts. "Did you just say you followed Brenda Matthews? Where— hang on—I just got your pin. What in the world are you doing out at Lake View Farms? I told you before Brenda Matthews is no killer. I hope you aren't getting into trouble because you are across the county line and out of my jurisdiction. I swear, Indigo, you are going to give me a—"

"Sean," I say. "I'm pulled off to the side of the drive, and two cars have just passed me. One has a yellow loofah, and I think I just saw Mayor Frank's car pass by with a blue one dangling from the open sunroof. All the cars out here have these different colored bath scrubbers attached. Something is going on—something weird. And what did you call this place, the Lake View subdivision? I don't even see a lake."

"Why would you? It's a new subdivision the Matthews represent, and I believe the lake is still under development. They are waiting on permits and such. But wait—did you say something about loofahs?" Sean asks, his voice is incredulous.

I hear a guffaw of laughter through the speaker. "Sean? Are you laughing? What's so funny?"

"I hope you don't have one of those spongy things dangling from your car's mirror, do you?" Sean asks.

"Of course, I don't. Why would I?"

"Somehow, Indie," Sean chortles, "you've managed to stumble onto a swinger party scene. And I strongly suggest if you don't plan on attending, you get yourself back to Corley County ASAP."

Swinger party? *Ew.* What the—Brenda and Frank Matthews? All of these people here are swingers? "How do you know that?"

"Haven't you read in the papers about all those senior citizen places in Florida? Swingers display different colored bath scrubbers to indicate their sexual preferences. I can text you the color chart if you want me to spell it out." Sean snickers. "Out of curiosity, what color did you say was on Brenda's car?"

"Senior citizens? Swingers? But Brenda and Frank aren't seniors. And I didn't say—but today, she has a black one on her car." I can almost visualize the self-satisfied smirk on Sean Riordan's face. "Is black among the naughty ones?"

"Boomers—what can I say? And yeah, a black scrubby means Brenda Matthews is a very naughty lady if you know what I mean?"

I'm afraid I do.

Feeling humiliated, I drive back toward town, avoiding Main Street altogether and the possibility of running into Sean Riordan. The echo of Sean's words and the glee he'd taken when he informed me that a black loofah indicates the desire for full swap burns in my ears.

Yuck! I thought such stories of baby boomers extolling the sexual prowess of the seventies were urban legends. I was mistaken. Once again, I ask myself just what kind of detective I am.

Chapter Eighteen

- -

A s penance for deserting my post in the shop, I volunteer to label a batch of
honey goat milk soaps. And if my self-imposed banishment to the kitchen
isn't enough, Sean Riordan pushes through the swinging doors. The exact person
I'd hoped to avoid.

"Hey, Indie, Heather said you were back here. I come bearing clean boots,"
Sean says. He holds up a pair of muck boots, immaculately clean. "Knowing how
attached you seem to these things, I figured you might be missing them by now."

"Wow! Spotless. Thank you." And thanks, Heather, for outing my hiding
place. Somehow, seeing my boots in pristine condition only makes me feel worse.
Pushing back my embarrassment, I put down the soap bar and turn to face him.
Sean's black eye patch is back in place today. "I see you managed to polish your
boots as well. They are shiny."

"Old habit from my military days," Sean says. "Always keep your socks and
your powder dry." Slowly, Sean flushes. "That sounded way worse than it did in
my head."

"Look, if you've come here to gloat, you can skip it. I feel humiliated enough.
Seriously, how was I supposed to know we have a group of swingers in Corely
County?"

"Correction, Douglas County, which is why I told you it's none of our
business."

"Okay, okay, out of our county," I agree. "But swingers—really? Ugh!"

"Hey, to each their own. It's not exactly illegal."

"Well, it should be!"

"What?" Sean glances down at the wristwatch. "Mark the time and date—for once, Indigo Evans and I agree on something. I'm a one-woman-at-a-time kind of guy. With this investigation demanding every minute of the day, I don't even have time for one."

"You're not seeing anyone?" Then, wishing I hadn't asked, I say, "I mean, not that it's any of my business."

"In case you haven't noticed, things have been a little crazy around here. A lot of it, thanks to you," Sean says.

"I—I'm sorry," I don't say the quiet part out loud. So not sorry.

"Indie, I didn't come here to pick a fight." Sean holds up the boots. "Truce?"

"Truce," I reply. "So, in the spirit of new-found cooperation, can you tell me what Josh had to say about dumping the ATI/EX 50/50?"

"You're not going to let go of this, are you?" Sean sighs.

"Nope." I point to my hair. "Stubborn runs in my family."

"I like the color of your hair," Sean says. "It reminds me of a burnished copper penny."

I stare at him. "Seriously? That's the best comparison you can come up with, the color of an old coin?"

He shrugs. "I told you it's been a while. Okay, how 'bout this? The color of your hair rivals the russet rays of a Kansas sunset. Better?"

I feel slightly mollified. "A little better, but you need to up your *A* game, Riordan."

"I hope to," Sean replies with a smile. "Just as soon as this case gets solved."

I remain utterly still as Sean reaches out his hand to trace the side of my cheek. He gently tucks a loose strand of hair behind my ear. His fingers feel warm on my skin and slightly rough, like someone used to doing hard work. An intense feeling of longing creeps into my heart, which surprises me. Then, like a typical man, Sean ruins the moment.

"It's adorable when your face turns all pink and freckled like a baby pig."

I slap his hand away. "A baby pig? Seriously?"

"What? Piglets are cute, all pink and wiggly and—"

"And they grow up to be bacon, which I do not eat, by the way!"

"Oh, yeah," Sean says, smiling. "Sorry—I guess that wasn't a great analogy. But, come on, Indie, you must admit, baby pigs are adorable. And bacon is delicious."

"At least the first part is true. But can't you give me a little hint of what Josh had to say? Was it a problem with the formula or an error on the production side? And please tell me this was just a one-off, and Agro-Tech hasn't been spraying deadly poison all over the countryside?"

"I'm pretty sure it was an isolated incident and so far contained to Eisner's property," Sean says. "But since it's going to come out sooner rather than later, I might as well be the one to tell you," Sean says. "We've determined that Agro-Tech recently received a couple of big orders for *Glo-Grow*. The plant added a third shift to keep up with demand.

"Josh went in for a couple of nights to cover staffing shortages. Perhaps it was in haste or ignorance, but he somehow reversed the formulation percentages while processing it. When he realized what happened, he panicked and ordered his crew to dump the contents into a ditch on the back of his uncle's property."

I groan. "Earl Blake? That property butts up against Eisner's and is practically a junkyard. The county is always after him to clean it up."

"Yeah, well, at this point, Uncle Earl has been issued a citation for illegal dumping and is none too happy about it. However, unbeknownst to Josh, the county had recently dug a runoff ditch to divert the water off 187th Street. I guess that part of the road is prone to flash flooding. With the dump site located uphill from the trenched area, the heavy rains from the other night washed the toxic stew into Eisner's watering hole."

"Oh my. Those poor raccoons. What a mess, and how stupid of Josh to think no one would notice. Perhaps he should consider a new line of work that doesn't allow him access to hazardous chemicals."

"Well, after today, his access will not be a problem. Carson let him go from his shift foreman position. Suspended without pay. Indefinitely."

"Yikes, I bet that was awkward."

"Probably so. But I have to give Carson Wells credit. He is taking full responsibility for the cleanup. He is liaising with the EPA and offered to resign once the cleanup is complete. Not sure how all that will shake out within the company."

"Well, I guess that's something. But we still don't know who killed Professor Fontenoy and attacked Karl."

"No, but we—as in me and the authorities—are handling things. You need to—"

"I know—I need to stay out of it."

"Indie?" Heather pokes her head into the kitchen. "Phoebe Sutter is here and would like a word."

"Phoebe Sutter? What does she want?" I couldn't think of anyone less likely to come calling on me. Especially voluntarily.

"I have no idea," Heather says, pointing at Sean. "Shall I send her back?"

Sean quirks his eyebrow. "I'd be interested in hearing what Phoebe Sutter says."

Of course, now our intrepid law enforcement officer wants to share sources. I make my displeasure known by glaring back at him. Boots, be damned. "Interim Sheriff Riordan just dropped off my boots. Wasn't that nice of him to clean them? But now he has to return to work, so send Phoebe in."

Phoebe pushes through the swinging doors. Spying Sean Riordan, she freezes in place. I'll never know how a man can emanate such authority while leaning against a stainless-steel worktable.

Truth be told, I harbor a certain amount of sympathy for Phoebe Sutter. This confined space has the look and feel of an ambush. However, catching Alex and Phoebe together in a compromising position still leaves a sour taste in my mouth. At least I no longer feel the deep pain of betrayal. In other circumstances, Phoebe and I might be friends. Yes, she is an eccentric, her intellect swallowed up in a goth-like persona. But I've been surrounded by quirky, creative types all my life, and I secretly wish I were free to be more like them. Indeed, I, of all people, know how charming and persuasive Alex Carmichael can be. I can't blame her for falling for his line of B.S. I did.

"Phoebe, you wanted to see me?"

"Ah—I—didn't realize you had someone back here. I can come back later," Phoebe says.

"Don't mind him. Interim Sheriff Riordan was just leaving." Sean doesn't respond immediately, remaining relaxed a beat too long. I realize his vibe is deceptively intimate and take a fair amount of satisfaction at the surprise that crosses Phoebe's face. But, finally, Sean stands upright and places the Stetson on his head.

"Ladies," Sean nods his head deferentially. "I'll leave you to it then." He indicates the spotlessly clean boots on the floor. "Indie… you know where to find me the next time you need your boots cleaned."

Phoebe and I fall silent as the bundle of masculine energy that is Sean Riordan leaves the room. I'm pretty sure Phoebe feels relief at his departure. Yet, somehow, I feel strangely empty in his absence. The man cleaned my boots. No one has ever done such a menial but meaningful task for me before.

"Indie? Did you hear me?" Phoebe asks.

Her question interrupts my musings. "Why are you here, Phoebe?"

"I wanted to know what happened to the *Lampropeltis triangulum gentilis*?" Phoebe asks.

"The what?" For a moment, I'm confused by her line of questioning. "Oh, you mean Bob, the milk snake?"

"Bob? Did you name the specimen Bob?" Phoebe's eyebrows raise in astonishment. "Seriously?"

"Yes, I did, seriously. I named the snake after poor Professor Fontenoy. I thought it to be a fitting tribute. Don't you?"

Phoebe's porcelain-like complexion turns chalky white as she stares at me in horror. What? Did she think I was kidding? Or is her corporeal reaction a result of a guilty conscience? Trying to prod for another incriminating remark, I persist. "I bet you thought releasing a snake into our apartment was amusing. Not many people know about my snake phobia."

"Alex told me," Phoebe blurts out, then quickly covers her mouth with her hand. "I'm sorry, I shouldn't have done that. It was a—prank. I've been worried sick. You didn't hurt him, did you?"

"Who? The snake or Alex—? No, of course, I didn't hurt the snake. Interim Sheriff Riordan and I released him on my father's farm. As far as hurting Alex, the jury is still out."

"Good, I'm glad. About Bob. So, are you and the new sheriff seeing each other?" Phoebe asks, glancing at the door Sean exited.

"No—no, of course not. Although, the circumstances of this investigation have managed to throw us into each other's paths a lot lately." Whether we liked it or not. Was it weird that I enjoyed being hurled, kicking and screaming into Sean Riordan's world?

"Well, that's too bad. The sheriff certainly fills out his uniform well," Phoebe says.

"Really? Well, not that he is mine for the taking, but don't you get any ideas. I think seducing Alex was more than enough. Anyone else would be overkill." Inwardly, I wince at the word choice.

"For what it's worth, I apologize," Phoebe interrupts. "And Alex was the one to make the first move. You know his ego can't resist a challenge." Phoebe covers her mouth again. "Not that you weren't enough of a challenge. Alex talked about you incessantly—to the point where I gave in to his advances just to prove I was smarter and prettier than you. But Alex felt your relationship had grown stale. I knew I was a diversion, but I got caught up in it and started to care about him. I'm not sure why. He is not my type. In hindsight, I probably did you a favor."

Stale. Ouch. "A favor? Is that what you call it?" But, in retrospect, Phoebe had done me a solid by opening my eyes to the one-sided relationship between Alex and me. In truth, I supported him throughout his Ph.D. program and encouraged him to take the position at Colette to advance his academic career, not mine. "Whoa—wait a minute—you said 'was'? What happened—did Alex cheat on you too?"

"No," Phoebe insists. "Nothing like that. It's just—"

"It's just what?" When Phoebe hesitates, I repeat. "What is it you want to say to me, Phoebe? Why exactly did you come here?"

"I'm worried Alex might be in over his head," Phoebe says. "He is going down the wrong path, so to speak. He's been acting strange ever since Professor Fontenoy was killed. I think he knows more than he told the cops."

"Strange in what way? Guilty strange?" I force myself to maintain a non-accusatory tone but am wary of Phoebe's openness. Who's to say she isn't the killer, trying to shift the blame onto Alex? That's what I would do, and Phoebe is an intelligent girl.

"Once the department received the Agro-Tech grant, Alex changed. He started dressing differently, drinking expensive wine, and going out to dinner with the higher-ups in administration. Did you know he sold his old Jeep to buy a black BMW? It's used, but don't you find it pretentious?"

"Alex was always pretentious. I just never minded—always envying those with the so-called finer things in life." Far more than his research, his pretension was one of the things I found both annoying and endearing about him.

"I know. At first, Alex's ambitious streak seemed kinda sexy. He could be so—well, you know. But now, he's different," Phoebe says.

"How so?"

"Right after Fontenoy's death, he acted entitled to the department chair position. I know, typical Alex. But, since your sheriff questioned him again, he's become strangely quiet. Almost reclusive. He spends most of his time in his office at the back of the lab," Phoebe says.

I chose not to remind Phoebe she spent most of her time in the lab, on her back, under Alex. There is no need to be spiteful. "In the first place, he is not, "my" Sheriff and Sean would never reveal a conversation with a source." Not that I wouldn't try to wheedle the information from the tightlipped law enforcement officer.

Phoebe muses. "Nothing has been the same with Alex since then. He is worried."

"Professor Fontenoy's death puts Alex's goal of tenure within his grasp, but maybe the idea of stepping over a dead body to do so weighs on him. But, hey, if you're so concerned about Alex's well-being, you should speak to Sheriff Riordan."

"Oh no, I don't want to get Alex in trouble," Phoebe says hurriedly.

"Phoebe, if Alex withholds information pertinent to this case, he could be in serious trouble. And you could be charged as an accessory."

A flush of color fills Phoebe's face. "You can't think I had anything to do with the professor's death? Fontenoy was always kind to me. This past semester,

I helped him verify some field study results on milkweed, looking for traces of clothianidin. He couldn't have been more grateful to have a tertiary verification."

"Milkweed? Phoebe, that was the study I performed, testing milkweed for residue. When I reported my results to Alex, he said the findings were incidental and of little consequence."

"Did you input your results into the mainframe of the department?" Phoebe asks.

"Yes, but the file was encrypted, not for shared use."

Phoebe's face flushes even deeper. "I might have helped Professor Fontenoy unencrypt several files. Fontenoy told me it was only for corroboration, so I didn't see the harm." Phoebe, realizing she's said too much, falls into silence.

"I knew it! Why didn't you speak up before, Phoebe?"

"It didn't hurt anything. The data was of incidental value, and the department needed the grant money. You know how tight the budget is for small universities, and I didn't want the loss of the grant to affect the accreditation review this fall," Phoebe says.

Like that was a valid excuse. Ugh! I had been right all along. Not only had my data been co-opted into the grant proposal, but the now-deceased department head had also accessed it. "When exactly did you help Professor Fontenoy access this information? Before or after the grant was awarded?"

"Oh—it was after. It was right before the end of term," Phoebe says.

"Phoebe? Did Alex happen to lose his faculty keycard recently?"

"Yes, how did you know? He misplaced his keycard a couple of days ago. He had to borrow mine to get into his office today," Phoebe stammers. "Why?"

"Come on, Phoebe. We need to speak to my mother immediately.

Chapter Nineteen

Phoebe and I burst into Oaxaca, my mother's retail space. The shop showcases Claire's hand-woven artisanal pieces and various responsibly sourced pieces and ceramics from the Zapotec people in the central valley of Oaxaca, Mexico. The sheer magnitude of the inventory and vibrancy of the colors is a feast to the eyes. Usually, I stop to finger the soft textures, but I walk past them today.

"Mom! I'm not sure you know Phoebe Sutter?"

"Indie—goodness, you startled me." Claire looks up from her weaving loom in the back of the shop. Acknowledging Phoebe, her voice hardens. "Yes, I know who she is—"

"I know this may be an odd question, but what can you tell me about the swingers' scene in the county?"

"What... swingers!" Claire exclaims. "Indie! Where did you hear such a thing?"

"Oh, come on, Mother. There isn't a bit of gossip in this county you haven't heard about. So, tell me what you know."

Claire clears her throat and cuts her eyes toward Phoebe, whose expression betrays a hint of puzzlement.

"I might have heard certain rumors, but I don't know any details. For certain, I've never attended one."

Well, that was a relief. "I'm more interested in the names of any participants. I know about Brenda Matthews, but are any Colette faculty members involved? Professor Fontenoy, perhaps?"

"Good heavens, Indie. This line of questioning is ridiculous. As for Brenda Matthews—I warned you she is willing to do just about anything, so it doesn't surprise me. But Bob Fontenoy—no way. I can't speak for all of the faculty, but I believe most of them are too afraid of Bob's ex-wife."

"Fontenoy's ex-wife. Who is that?"

"Why, Victoria Medford, of course," Claire responds.

Ex-wife? Once again, my mother had buried the lead. Ugh!

"Professor Fontenoy and Dr. Medford were married?"

"Oh yes, but that was years ago. They were grad students together at Michigan," Claire says.

"Michigan, seriously?" The prestigious school in Ann Arbor is a top-tier school for science. "How in the world did they end up here?"

"I don't know all the details other than they were married young, and each of their career trajectories took a different path—as things are wont to do between married academics. Poor Bob—well, his star was completely eclipsed by Victoria's, so to speak."

And yet they both ended up here. Quite a coincidence for someone who did not believe in them.

"Did you know any of this?" I turn to Phoebe.

"No, most definitely not," Phoebe says.

"Phoebe, do you trust me?"

Phoebe hesitates. "Yes—yes, I do."

"Good. You must go to Interim Sheriff Riordan's office and tell him about your concerns. Alex could be in danger. And while you're there, see if he knows that Victoria Medford was once married to Professor Fontenoy."

Phoebe looks between Claire and me. She shakes her head in the affirmative and hurriedly leaves the room.

I turn back to my mother. "Claire, I need your faculty keycard."

"I can't give you that, Indie," Claire says.

"Why not?"

"Because I don't have it. I gave it to Giovanni, and he's working in the studio today in the Art Department," Claire explains.

Geez. Talk about lax security standards. "Call him, Mother, and have him prop open the side entrance door."

Giovanni has wedged a chunk of granite between the metal door and the frame. Lucky for me, security is nowhere to be found. I know from experience the elevator in the art department is out of order, so taking the stairs is my only option. The building's elevator, a sixties-era, minimalist-style structure, is always broken. Back in the day, the architecture was deemed transformational. Today, the style is passé with the construction of a more environmentally friendly version to begin next year.

Slightly out of breath, I burst into the third-floor studio space. A patina of stone dust floats in the air, making me sneeze. The windows are wide open. Even so, there is only a minimal breeze, and it's unbelievably warm inside. Alerted by my sneeze, Giovanni looks up from his work. He is shirtless, his muscular upper torso coated in a white powdery film.

Giovanni puts down his chisel and removes a pair of eye goggles and a protective mask covering the lower half of his face. "*Cara?* What is wrong? Your mother sounded frantic on the phone." Giovanni grabs the white T-shirt hanging nearby and dabs his face, then his chest, before slipping it on over his head

For the first time, I notice the burning intelligence in Giovanni's soulful, dark eyes. The man is sinfully handsome, and I can see how easily Claire could fall for him. I mean—who wouldn't? There is also an innate kindness in his eyes that perhaps my mother gravitated toward—something she'd been missing from her life, especially from me. I remember all the snide remarks and innuendos I'd cast her way this past year. Today, I feel a sense of shame and resolve to make up to her. Somehow. Had my father been unfaithful? Someday soon, I would ask, but today was not the day. I'd always put my father on a pedestal, ranking him morally superior to my mother, and yet she never disabused me of the notion.

"Giovanni, for what's it worth, you and my mother have my blessing. Life is short, and if I've learned anything these past few days, one shouldn't be afraid to take chances. It doesn't matter what I or others think. I want my mother to be happy."

Giovanni smiles; his face lights with pleasure. "There is a quote from Dante—I believe the English translation is, *'Follow your path and let the people talk.'* Perhaps now is the time to follow your path, Indigo Bleu."

"Follow your path," I repeat, mulling over the words. Now is the time to follow my path.

Eureka. "I know who did it."

"What? You know who did what?" Giovanni asks, genuine confusion evident in his tone.

"I know who killed Professor Fontenoy! I gotta run. But you better be good to Claire, or you'll have me to deal with. *Ciao!*"

"No, *Cara*—I would never— But wait Indie, perhaps you should notify the *polizia*. Perhaps this knowledge is dangerous?" I hear Giovanni call out behind me, but it's too late. I bolt from the studio at a full sprint.

I dash across the quad, my long legs eating up the short distance in no time. Giovanni's words echo in my head. "Follow your path." It was obvious. My path led straight back to where it all started, the science department building, The Mudge.

Using my mother's faculty keycard, I slip into the building unnoticed. Passing the abandoned security kiosk, I wonder where security is. Out making rounds? All the better. As a former student, my presence would require an explanation—one I could not give.

During the summer, most university buildings are deserted at this time of day. I take the stairwell to the basement level, which houses the majority of the science labs. The theory being volatile chemical compounds are more stable when kept at consistent temperatures. The same rationale applies to the location of the entomology lab. Insects and reptiles thrive in cooler environments. Plus, the administration is keen on keeping potential escapees below ground.

"Abandon all hope, ye who enter here." The only quote from Dante I recall from freshman literature classes comes to mind for some reason. I descend the darkened stairwell and follow the long corridor to the entomology lab. I feel as if I'm crossing the River Styx. The hallway temperature is frigid, and I immediately regret the lack of a lab coat. Perspiration dries on my skin, producing chill bumps, while the dark hall triggers a flood of images, one of late afternoon the prior semester.

I remember thinking there must be some mistake. But no, the email from the thesis committee was explicit. *Upon preliminary review, Dr. Robert Fontenoy, Department Chair and doctoral committee chair, rejects the analysis of glyphosate and its effect on the gut microbiota of the honeybee.*

In the strongest possible words, the committee warned me against the ethics of plagiarizing data. As it turned out, data identical to mine showed up in the Agro-Tech Industries, LLC grant application filed under the auspices of the university's science department—and Dr. Alex Carmichael. Coincidence? I think not.

Once again, I fight back the rising tide of anger. I am sure Alex incorporated my data into the grant submission packet. I frequently stayed overnight in his apartment, and he had access to all my credentials. How could he do such a thing? At the very least, I had believed him to have too much scientific integrity. Obviously, I was mistaken. Worst of all, I thought he loved me, but you don't betray the one you love. Our relationship was nothing more than a lie.

With a strong sense of *déjà vu,* I scan the key card. The light flashes green, and the door clicks open. It is a standard college lab, cluttered and chronically under budget. Long, black, epoxy resin countertops, nonporous and flame retardant, add an overall gloom to the large room. Glass-fronted terrariums line the walls, requiring only ambient light to keep their inhabitants, mainly of the order *Blattodea*, which includes cockroaches and termites, alive.

This lab is designed for first-year entomology students. As an associate professor of the science department, Alex and Phoebe, his graduate assistant, taught the course. Now, with the untimely death of Prof. Fontenoy, Alex was catapulted into the position of interim department chair. His academic prospects had undoubtedly improved over the past few days. In hindsight, those circumstances should have made him, not me, the chief suspect.

I wind past twin terrariums lit up by the red glow of infrared heat lamps and bypass the one housing a female Texas brown tarantula, fondly known as Nefertiti. I avoid altogether the lab section holding the glass-fronted aquariums of nonvenomous snakes. Thankfully, the university had placed a moratorium on venomous snakes following an incident where a mature male copperhead was loose in the lab for days. Still, I'm careful to go nowhere near them. Instead, I head to Alex's closet-sized office at the back of the lab. A tiny sliver of light is visible beneath the office door. Maybe Alex is there after all. Good. I would resolve this issue with him once and for all.

Recalling what happened the last time I'd hastily thrown open the door without knocking, I rap my knuckles sharply against the wood. Although, my feelings are strangely ambivalent at the thought of finding Alex embracing Phoebe or anyone else.

"Alex—are you in there?" I turn the knob and push against the door, half expecting a reply. The door opens a few inches and then resists. "Alex?" Using my shoulder, I shove. It opens a few inches more. Something is blocking the door from fully opening.

I push to the point where I can wrangle my arm through the crack and switch on the overhead light. I whisper Alex's name. My previous outrage has dissolved into something akin to fear and trepidation. Pressing my eye to the gap, I glimpse the heel of one brown leather shoe sticking out from behind the edge. It is Italian. No socks. A sliver of pale flesh is visible above the heel. Alex? I begin to hyperventilate, and a guttural moan escapes my lips. NO—oh, no!

Forgetting about the lack of service within the thick walls of the basement level, I reach for my cell phone. What to do? Should I force my way inside, possibly further injuring Alex? Or is it already too late? God, I pray not. Should I run back and try to find the missing security guard? Phoebe's words ring in my ears as I shimmy through the space.

Phoebe had warned me Alex was in over his head. I mistakenly believed she was trying to implicate him in Fontenoy's death. But had Alex's ambitions caused his death? No one should die for a dream.

At last, I wiggle my body through the gap. Alex lies face down on the cracked linoleum floor. A familiar, eerie stillness envelops me as my stomach turns to ice water. It's too late. So, I push back the immediate sense of loss and bend down to pick up his wrist. The skin feels cold and stiff to the touch. There is no pulse. Trying desperately to control my breathing, I step back to view the body as dispassionately as possible. The time for mourning will come later, and despite everything Alex put me through, I will mourn him.

Squatting down next to the body, I balance on the toes of my running shoes, careful not to let my bare legs touch the floor and further contaminate the scene. I should check for a carotid pulse, but seeing the dark red pool spreading beneath Alex's head, I cannot force myself to touch his neck.

I search for the landline, *de rigueur* for all basement-level rooms, and notice the slag glass lamp, broken on the floor. I remember the day Alex and I had picked up the reproduction at an estate sale in Madison for a song. Alex loved that lamp, saying it added a touch of class to this otherwise utilitarian space. The shards of antique glass lay in a colorful pattern across the faux oriental rug beneath his desk—an additional personal touch for the newly minted associate professor. Amazingly, the lamp's heavy bronze base remains plugged in. The glow of the incandescent bulb illuminates a splatter of dark red on the floor. Is that blood? I step carefully over the glass fragments to peer closely down at the base. Could this possibly be the murder weapon?

"Stop where you are! Put your hands in the air and turn around nice and slow!"

I comply and turn to find the absent security guard in the open doorway. His sudden arrival is one mystery solved. But I wince, noticing the lower half of Alex's body lies in a cockeyed position, a casualty of the guard's haste upon entering the room. The young man is ghastly pale in the glare of the overhead fluorescent lighting. However, he holds a lethal-looking black handgun in a two-fisted grip. It's pointed straight toward me. Noticing my sideways glance, he looks down. The gun in his hand begins to tremble.

"Jesus—what have you done?"

"I—I didn't—" I gesture toward Alex's body with one hand. "I—"

"Put your hands back up. Don't move, or I'll—"

"Take it easy, officer, and lower your weapon. I'm Interim Sheriff Riordan from the Colette County Sheriff's Department," Sean says. Standing in the doorway, he holds his badge in one hand, and the other rests on the butt of his gun.

The wide-eyed guard swings around, and his weapon rises skyward. But before I give a warning, Sean draws his sidearm and points it toward the startled man. I notice his grip is firm and steady.

"I said lower your weapon, officer. That's an order!"

Something in Sean's sharp command must penetrate the guard's bundle of skittering neural synapses, and his gun hand drops.

"Geez, don't scare me like that. I could have shot you, Sheriff. I was out making my rounds, checking the lab when I saw this door ajar. I came to investigate. I found her and—him. Is he—" he gestures toward Alex with the tip of his gun.

"Sean, Alex is dead." I interrupt the rattled guard. "I found him lying on the floor, wedged behind the door." My voice breaks. "I think someone bashed him over the head like Karl and Professor Fontenoy. There's blood—and, oh—I think I know who did it."

"Who?" The guard says incredulously. "She's the who. She was standing over there with that lamp, probably the murder weapon. You should arrest her, Sheriff."

"Okay, let's all slow down and take a deep breath. Officer—? I'm asking you to holster your weapon for everyone's safety."

Sean does not lower his gun. Instead, he maintains an acute awareness of the current threat: the flustered campus security guard holding an unsecured weapon.

"Cates—Brandon Cates." The young man looks down at his gun as if seeing it for the first time. Yeah." He slides the weapon back into his side holster. "It's the first time I've drawn my weapon on duty. But, since the murder of the other Professor and now—–"

Relieved, I expel the breath I hadn't realized I was holding. My emotions are pinging all over the place. "Sean—" I start.

"Like I said," Officer Cates continues, "I caught her red-handed standing over the victim with the murder weapon."

"That's not true!" I retort. "Anyway, how do you know—?"

"That's enough, both of you!" Sean barks. With his weapon leveled in one hand, he squats down to check for a pulse on the left side of Alex's neck, never taking his eye off either of us. Finding the lack of it, he grimaces, stands upright, and holsters his weapon. "Officer Cates, did you not bother to check the victim? Because I'm sure if you did, you would notice the blood on the back of the professor's head appears to be from an exit wound. I suspect we will find an entry wound on the front of his face when we roll him over. The assailant fired from point-blank range at some point earlier in the day, judging by the amount of the body's rigor. So, Officer Cates, where's the smoking gun if Ms. Evans is the perpetrator?"

Alex—shot? I inadvertently made a mewling sound as both men looked toward me.

"Well, I—I mean, she was standing over there by the desk—looking at the lamp and," Officer Cates blusters, "so I just assumed she—?"

"Never assume, Officer Cates," Sean responds. "Your job is to assess the situation, contain any possible threat, and prevent further escalation of hostilities. Ms. Evans is not the threat. At best, she is a witness after the fact."

"But—how could I know?"

"Well, now you do. Besides, an eyewitness puts Ms. Evans' arrival in this building approximately—" Sean looks at his wristwatch. "—ten minutes ago. We can safely assume Professor Carmichael has been dead for over ten minutes. Did you hear a gunshot any time before that? And how did Ms. Evans slip into the building undetected?"

Now, Officer Cates blushes furiously. "I—I stepped out to use the john next door in the Business School. "It's just me—and well—the one in here was clogged up, and I don't get off 'til midnight," he mutters.

"Hey, I get it. Things happen. But now this has become a murder investigation, and I'm taking charge of the scene. Do you understand? I need you to go upstairs and wait for backup from the Sheriff's Department. Send my deputy down immediately and tell them to notify the coroner ASAP."

Warily, Cates eyes me. He nods his head in the affirmative. On his way out of the office, he leaps over one of Alex's legs, which lies at an awkward angle. Annoyed, Sean shakes his head but offers no further comment. He doesn't speak at all.

"Sean?"

He holds a hand up, "*Shush*—Indie, please, for once since I met you, just answer my questions. Did you touch anything else in this room?"

"No, I knocked on the door and then turned the knob. I checked for a pulse on Alex's wrist. But—shot? Are you sure? I thought someone hit him in the back of the head like Karl. So, when I noticed the broken lamp, I came over to examine it for evidence of blood on the base. But, I swear, I didn't touch it. Oh God—do you think someone shot Alex? Really?"

"Yes, I do. And, since the crime scene is compromised, you and I will carefully step back out the door. Follow your exact path back from the lamp."

Path. There was that word again. "Sean—"

"Out," Sean orders. "And then perhaps you can explain why I got a call from a half-crazed sculptor. He was shouting in a mixture of Italian and English something about you running pell-mell across the quad, then entering a building you're no longer authorized to be in?"

"I know—I know, it was stupid." Stepping carefully through the doorway, I refuse to look down at Alex's body. "I still can't believe Alex is dead. And here I was worried he was the guilty party when Alex was in danger all along."

"You can't blame yourself, Indie," Sean says. "The blame lies with the killer."

"Phoebe tried to warn me."

With a face void of expression, Sean asks, "What exactly did Phoebe tell you?"

"I don't know—everything is mixed up in my head right now. I mean, originally, I suspected Alex or Phoebe killed Fontenoy to advance Alex's prospects. But now—?"

Exasperated, Sean sighs. "Was this before or after you were convinced the killer was Carson Wells? Only then to discover he was covering for a mistake made by his lover, Josh Blake—your high school friend, everyone in the county, knew, except for you, was gay. And then—let me see—who was next on your little suspect list? Oh yes, Brenda Matthews, the mayor's wife. You just had to expose her involvement in some secret swingers' club. Every time I see her, I cross to the opposite side of the street to avoid her. Because I know she knows, and I can't look her in the face." Sean pinches the bridge of his nose and closes his right eye. "This is what happens when civilians try and play Nancy Drew."

"Well, to be fair, you suspected Wyatt." Hardly a valid excuse, but, hey—I tried.

"Don't even go there," Sean growls. "I get a headache trying to figure out Karl–Wyatt's genealogy."

"Okay, I admit I was wrong about Wells and Brenda. And Wyatt's secret turned out to be a good thing. Karl has gained a grandson. But today, Phoebe said Alex was traveling down the wrong path—the wrong *path*! Don't you get it? Phoebe was referring to the cow path to Eisner's pond, contaminated with ATI/EX 50/50 runoff. Did she come to you?"

"Yes, Phoebe told me she was concerned about the 'path,'" Sean uses his fingers to form quotation marks, "her increasingly erratic ex-boyfriend was going down." Sean slaps one palm down on his thigh. The sound echoes in the room, causing the chirping crickets to go silent. "And as it turns out, her concerns were well-founded. I cannot believe two such intelligent women as you and Phoebe Sutter could have fallen for the line of BS from the now-deceased Alex Carmichael."

"Did Phoebe tell you about Fontenoy and Dr. Medford?"

"Victoria Medford? Yes, I knew, and I've got an APB out on her whereabouts, but so far, we haven't been able to locate her," Sean says. "Frankly, I've been too busy chasing after you. And now we have another murder to deal with."

Suddenly, it all clicks. How had it taken me so long to figure this out? That's what Phoebe had meant about going down the wrong path. "Sean, we have to find Phoebe right away. She could be in danger. I think Victoria Medford is the killer."

Chapter Twenty

- -

Sean and I both turn as we hear the lab door open and watch a flashlight beam sweep across the room.

"Riordan?"

"Yeah, Nolan, we're back here," Sean replies.

"Geez, can we turn the lights on? This place is creeping me out."

"Go ahead. But be careful where you walk; there could be blood splatter, and you need to watch for shell casings. Is the coroner on the way?"

The overhead lighting illuminates the room, and I observe Nolan navigate the space. Finally, he reaches us and rubs his arms with his hands.

"It's cold here, and I hate snakes," Deputy Nolan grumbles. "Doc Barnes was none too happy about being interrupted on the back nine. He was sitting at two below par. By the way, Sheriff Kramer is on his way down here."

"Kramer?" Sean says. "I thought he was still on crutches?"

"He is, but he said he's coming even if they have to carry him in on a litter. He kept mumbling about how the county is going to hell in a handbasket," Nolan replies, a note of censure in his voice.

"Oh great, that's all I need," Sean says.

Wisely, I keep my mouth shut and shrink back against the wall. I need to locate Phoebe as soon as possible, but I bide my time while trying to escape Sean's watchful gaze. Nolan whistles as he peers through the doorway, then turns his baleful glare onto me.

"Oh man, maybe we should have locked you up after all," Nolan mutters.

"Don't look at me. I found Alex like this," I say defensively. But I am starting to wonder if Deputy Nolan and the rest of the department might be right. Had I become a harbinger of death with the proverbial black cloud hovering above my head? Maybe I should have taken Bea's analysis of my aura more seriously.

"Hernandez contacted the KBI, and the crime scene team is rolling this way. I left him upstairs with the Italian guy who kept trying to get down the stairs to check on her."

Nolan thumbs his big finger in my direction.

"Hernandez had to restrain him. I can't understand half of what he is saying, but he seems pretty determined, and boy, howdy, does he ever have a grip on him."

I blurt out, "He's a sculptor. He has strong hands." Both men look at me strangely. Oops. But my heart warms towards Giovanni. At least someone is concerned about me.

Sean stretches his neck from side to side, eliciting an audible pop. "Yeah, she seems to have that effect on men. Where's Officer Cates?"

"I let him run to the john. He kept moaning about eating a bad burrito and clutching his gut. Seeing your first homicide up close always gets to you."

"The first one is always the worst. Cates has got it together now," Hernandez says upon his arrival. "Sheriff Kramer is outside in his POV and wants you to come upstairs and brief him as soon as you can break free." Hernandez points directly at Sean.

"Doc Barnes here?" Sean asks.

"On his way down," Hernandez replies.

Deputies Nolan and Hernandez quickly cordon off the crime scene with yellow tape, official and CSI-like. Heather would be impressed. Doc Barnes, a semi-retired physician, arrives. When he was appointed county coroner, it was with the understanding he wouldn't be called to many crime scenes. He second-guesses his decision to accept the position as he slips a pair of *Tyvek* coveralls and paper booties over his golf clothes while muttering about "being too old for this." Honestly, I can't disagree. I'm not even thirty, but the past few days have me feeling ancient.

The coroner's arrival allows me to sneak out of the overcrowded with crime scene techs milling about. I slink toward the egress door installed to bring the building up to code a few years ago. It leads up a set of concrete steps into the faculty parking lot.

The lot is deserted except for one lone vehicle, a sleek, black four-series BMW coupe. It must be Alex's new ride. The car is undoubtedly a step up from the battered old Jeep Cherokee he'd driven while we were dating. It's a shame he'll never enjoy it. When you add the car plus the recent wardrobe upgrade, you get a man promised a reward financially as well as in academic ranking. Tears flood my eyes, and I force myself to shake off the temptation to wallow in my grief. Regrets will come, but my focus is on finding Phoebe Sutter alive.

With growing police activity at the front of the building, I stick to the shadows of the adjacent *Walter H. Beech School of Business*. I don't know much about the Beech family other than that Walter and his wife, Olive, founded Beechcraft Aviation in Wichita, KS, in the 1930s. The building's name is yet another example of the male patriarchy that dominated the period. The slogan, *"The future is female,"* still struggles to be heard. However, I feel pretty empowered as I avoid the various law enforcement types, including the burrito-eating Officer Cates.

Jogging quietly through the quad, I head to my parked car at the back of the Art Department. At one point, I think I hear someone shout my name, but I ignore it and drive beneath the university's arched stone entrance inscribed with the Latin words *Omnia grata sunt*, meaning *"All are welcome."* The motto is a gracious reminder of the mission of Colette of Corbie, the foundress of the Poor Clares, who I like to believe was an example of an early feminist. Without hesitating, I point my Subaru east and head toward Eisner's pond.

Under the 1976 Resource Conservation and Recovery Act (RCRA), a framework for the adequate disposal of hazardous waste was established. As appropriate for a pending cleanup, a three-rail traffic barricade blocks the dirt cow path, and a bright yellow hazardous waste sign is attached to the top rail. Ignoring the signage, I park the car and continue on foot. Only a few streaks of orange and gold penetrate the thick copse of trees surrounding the cow pond. Fifty yards in, the path narrows. I notice the tracks of a smaller vehicle veer suddenly off to one side. Upon further approach, I make out the outline of Phoebe's black Mini Cooper. The white racing stripes across the roof are barely visible in the waning light. Simultaneously, the mix of dead vegetation and strong chemicals hits my olfactory senses. *Pee-ew.* I'm definitely in the right place and quicken my pace.

A shrill scream echoes against the backdrop of the quiet evening. I stop dead in my tracks, trying to ascertain the direction of the sound.

"No—no, please—you don't have to do this!"

Phoebe! I recognize the voice and run headlong through the underbrush, heedless of the spiny bushes scratching my bare arms and legs. I stumble on an exposed root and crash through a gap in the foliage. Pain shoots up my left ankle, and I half limp, half skip, my forward momentum stymied. I almost fall flat on my face as a figure dressed in a protective beekeeping suit looming over Phoebe comes into view. Phoebe kneels in the damp mud of the water's edge, her hands raised to cover her face and neck. The white-garbed figure holds a box I recognize as a lightweight cardboard trap similar to the ones Karl uses to capture swarms.

"Please, I swear I won't tell anyone. I'm violently allergic to bees," Phoebe pleads.

"You fool! It's too late for that."

I recognize the voice of the university's president, Victoria Medford, and hear the buzz of bees within the confines of the white box. They sound pissed off.

Horrified, I watch Dr. Medford shake the box, further aggravating the bees inside. She fumbles with the thick cardboard lid, the leather beekeeping gloves encumbering her movements.

I don't know if the rush of adrenaline surging through me is from the pain in my ankle or the delayed reaction to the sight of Alex's lifeless body. But all my focus, and yes, I admit, all my rage, is concentrated on the woman raising the box of vibrating bees over Phoebe's head.

"Do NOT hurt my bees!" I shout and lunge forward, my pent-up frustration directed onto this person who has more than likely taken the lives of both Alex Carmichael and Bob Fontenoy.

Startled, Dr. Medford pivots, her feet slipping in the mud. The lightweight box flies from her hands, and a swarm of a couple of thousand angry bees escape from the open top and fly into the surrounding air. Medford windmills her arms in a vain attempt to regain her balance, further agitating them. They buzz angrily about her veiled face and obscure her vision. With a final scream, she falls backward into the muck, tripped by the deep imprints of cow hooves found on the edge of all farm ponds. Her backside hits the muddy water with a splash. How poetic to be felled by a domestic bovine. Take that, you psycho!

Meanwhile, my jump resembles the wobbly leap of a three-legged bullfrog, but I hit the target, toppling poor Phoebe. The weight of my body pushes her back into the mud. Somehow, I end up with my legs straddling her waist while my chest smothers her face.

With an *oomph* of exhaled air, Phoebe coughs and pushes against me with her hands. "Stay still," I tell her, my voice sounding hoarse from the impact of the fall. "The bees will calm down in a second. I'm trying to protect you. I had no idea you were allergic to bees."

"I'm—I'm not—" Phoebe sputters, her voice muffled.

"What?" I ask, raising my body a few inches away from her face. "What did you say?"

"I said I'm not allergic to bees. I just told Medford that to buy time," Phoebe says. "Now, get off me, you stupid sow."

What do they say about no good deed going unpunished? For the record, sows are not stupid; they are fierce. And truisms are usually valid for a reason. My rescue of Phoebe Sutter is proof positive.

A shrill catcall fills the air. I turn my head to see Chief Deputy Harlan Pierce standing less than six feet from us. He points a cell phone in our direction.

Click. "*Woohoo*—chick-on-chick mud wrestling. Hold still, DB. The guys back at the station will love this one," Harlan says.

I blink as the camera flashes several times near my face, and somewhere behind me, I hear the sucking sound of Deputies Nolan and Hernandez pulling a screaming Victoria Medford from the pond. As my vision clears, a hand, unsullied by mud or muck, extends into my field of sight.

"Need a hand, Indigo Bleu?" Sean Riordan asks.

Is it just me, or do I detect a note of sarcasm in his tone?

I wish I could tie this little vignette up in a box with a neat bow and tell you no one sustained further injury. Alas, I cannot. Honeybees do not discriminate and singularly focus their wrath only on the guilty.

Thankfully, most of the swarm had settled on a low-hanging branch over the pond, but a few dozen angry guard bees stung the rest of us. I got popped several times on the back of my thighs. Sean's left cheekbone sported a string of angry red welts. Nolan and Hernandez stayed busy swatting away kamikaze bees while dragging a resistant Victoria Medford through the mud. She struggled against them, laughing manically until Hernandez finally had enough. He reached up and snatched the bee veil from atop her head. I knew I liked him best. Hands shackled, Medford unleashed a profanity-laden tirade on the deputies, us, and the rest of the world.

Somehow, Chief Deputy Harlan Pierce managed to escape unscathed. He abandoned his quest for souvenir photos to assist Phoebe from the ground. Who, by the way, was covered from head to toe in mud, thanks to me. How's that for an aura color?

Anyhoo, we were a sad and dirty lot as we split up to head back to Colette. The subdued Victoria Medford is restrained in the back of Nolan and Hernandez's patrol car. Harlan focused his attention on the beestings circling Phoebe's bicep. The head of the serpent tattoo had rapidly become red and swollen. Harlan popped the clutch on Phoebe's Mini and rolled it out of the bushes, and Phoebe readily accepted Harlan's offer to drive her back to town. What was that about?

Interim Sheriff Sean Riordan spent his time talking into his shoulder mic, hopped into his county SUV, and roared down the cow path without a backward glance. Yeah, I would say he was pretty pissed at me.

I slide into my trusty steed—*ahem*—Subaru, thankful I had the foresight to throw in a couple of old towels and head home alone. As an aside, I do not recommend sitting on leather seats with bee stings on the back of bare thighs. As Mustard would say, "Yeow!"

The sun has set as I pass the entrance to Colette University. Seeing the flashing strobes of emergency vehicles reminds me of Alex Carmichael. His ambition was his undoing, but all of my previous animosity has dissolved into a profound sense of sadness and remorse. Life can be funny that way. Sometimes, a life's absence makes you appreciate what is essential in your own. Home, family, and yes, with all of its ups and downs and crazy emotions, even love.

The front entrance of The Hive is lit up as I walk back toward our shop. I'm dirty and exhausted, but the familiar voices of my family and friends, both old and new, beckon me.

Claire looks up and exclaims. "Oh, my—Indie! You're filthy!"

"Indie, my God, are you okay?"

I look up to see my dad standing there, looking concerned. "Hey, Dad, you're back! Glad you made it in time for the—ah—party?" I gesture at the little group.

"I—*ah*—" Daniel Evans says. "Your mother said you needed me, so I returned early." He clears his throat self-consciously. "And here I find you chasing down a killer. Indie, what were you thinking?"

"It's all good, Dad." The words to the song, *"I get by with a little help from my friends,"* come to mind. Truth be known, how boring life be without them? Surrounded by familiar faces, I feel a sense of contentment. My mother covers her mouth with her hands. Heather runs from behind the counter to throw her arms around me.

"Oh, Indie, thank God you are okay! We were so worried." Heather says. I wrap my arms tightly around my sister, mud and all.

My parents move to embrace Heather and me, and all my resentment and anger fade away. I'm content to bask in the love of my family. Giovanni, Wyatt, and Bea stand by, watching us.

Giovanni begins to gesture with his hands. *"La famiglia!"* His accent is thick. "How do you say in English—is it time for a group hug?"

Epilogue

A fter the craziness of the past few days, a good night's sleep goes a long way in restoring my good spirits. So much so taking out the garbage feels like a much-needed return to normalcy. I slam the dumpster's lid down with a little extra gusto.

Bob Fontenoy had been a scientist in the truest sense of the word. Worried his ex-wife had once again rushed the grant approval process, he'd performed a field study using the honey from my hives to test for glyphosate levels. Victoria Medford's history of fudging data to garner a lucrative grant and her general lack of scientific ethics led to the couple's divorce all those years ago. Dr. Medford's last employer had quietly encouraged her to seek employment in a less visible academic setting, so she jumped at the opportunity to head a small university in Kansas when the position opened.

Was it a coincidence the job happened to be at the same university where her ex-husband was chair of the science department? We'll probably never know. I would like to believe Bob Fontenoy had reservations, but if so, he did not voice them, and Victoria got the job.

Sheesh—talk about your classic example of failing upward. However, at some point, Bob Fontenoy confronted his ex-wife, and like a classic narcissist, Victoria refused to own up to her deception. Rather, she'd followed him to the orchard that fateful morning to continue the argument, which ended after Medford picked up a brick weight from the top of the hive and whacked the professor in the back of the head. Whether this action was premeditated or spontaneous will be up to the judicial system to decide.

Alex Carmichael was willing to participate in Medford's scheme by tweaking the grant data in favor of Agro-Tech's product. Although his ambitions overrode scientific ethics, Alex would never condone murder. After Fontenoy's death, he must have gotten cold feet and threatened to divulge his role in the skewed research proposal.

Medford's initial attempt to shift the blame onto Alex did not fare well. Her stealing honey from Karl's bee room was a deliberate attempt to throw suspicion onto Alex by leaving behind his lanyard and faculty badge. But, by some twist of fate, the key card itself was lost. Granny Bea would tell you the disruptive forces of a transit Mercury and a sextile Mars were the likely culprits. Me? I believe Victoria Medford was no criminal mastermind but I do respect the idea of karmic justice. Geez… maybe I am becoming a believer in such things? It turns out coshing Karl over the head was the lesser of Medford's evils. She'd become increasingly erratic in a desperate attempt to cover up her crimes. Alex ended up paying the ultimate price with his life. It all came down to greed.

Thankfully, the Agro-Tech spill was confined to Eisner's property. The EPA sent a cleanup team composed of several environmental consultants. I have yet to meet them. Hopefully, a state entomologist will be on the team to assess long-term damage to native pollinators. Agro-Tech has agreed to pay for the cleanup. But Carson Wells—well, he's been quietly transferred. I feel bad about that. It seems his culpability was fueled by ignorance and love. However, since lawsuits are pending, the company had to make a choice. My friend Josh Blake is no longer employed in the chemical compounding business, but he followed his heart and moved with Carson. I wish them both well.

The nucleus hives in Mueller's orchard continue to thrive, but who knows why one of my new package bees absconded? I guess that's a mystery for another day.

Out of the corner of my eye, I notice the lights of a Corley County Sheriff's Department vehicle pull into the alley behind me. *Whop, whop.* The blue lights flash. Seriously? What now?

Sean Riordan steps from the SUV dressed in the standard departmental khaki shirt and jeans. His badge is clipped to his gun belt, his sidearm holstered. "Is there a problem, Sheriff Riordan?" I raise my eyebrows in mock query.

Sean smiles. "Yes, ma'am, I've been getting a few complaints about the vegan cookies sold in your facility."

"Oh yeah," I say. "And what kind of complaints might those be?"

"Well, someone claims the cookies are made without eggs or butter. I ask you—is that even possible? Another one grumbled about the excessively crunchy cookies, claiming they almost broke a tooth.

"Well, don't plan on suing because I know plenty of good lawyers." Sean rapidly closes the space between us, and I'm forced to backpedal until my butt hits the building's brick wall. "As you might surmise, vegan cookies contain no animal byproducts. That would include both eggs and butter."

I press my palms into his shirtfront. *Hmm*—no ballistic vest today, just lean, hard muscle. "The extra crunch comes from aquafaba, and Angela uses a low-moisture sugar in the recipe. Both act as thickening agents."

"Aquafaba, what is that?"

The open palms of Sean's hands press into the bricks on either side of my head, effectively pinning me against the wall. From the amount of fluttering in my chest, I probably resemble a taxonomy specimen from the order *Lepidoptera*, otherwise known as the common butterfly.

"I'd love to know," he whispers.

His voice is low and gravelly, and as he leans forward to speak into my left ear, I feel the urge to tip my head toward his. The faintest hint of patchouli wafts through the air between us. "Aquafaba is the liquid drained from chickpeas." My mouth is suddenly dry. "It's used as a substitute for eggs in vegan baking.

"Hmm—chickpea juice. Sounds—yummy?" Sean says. "Kinda like you."

I shiver despite the warmth of the morning sun streaming into the alley.

He leans in further and whispers. "You owe me."

"I owe you?" Is this a dream, or is Sean Riordan flirting with me? Here? Outside, in broad daylight? If so, don't wake me up. "Why do you think I owe you?"

"For saving your life," he says.

"You saved my life? Of all the—" Nope, it's not a dream. "More like you owe me."

"Oh yeah—what for?" Sean queries.

"For solving Professor Fontenoy's murder, of course." And then, because I can't help myself and for some reason, I revert to total idiocy when I'm around Sean Riordan. "You should be the one thanking me for doing your job."

Sean's hands fall away to rest on his hips. He appears aware of our surroundings suddenly. The expression on his face becomes stony. Once again, he's not wearing the black eye patch. The bright blue orb of his custom glass eye glows soft and blue in the morning sun, a web of faint scars visible around the eyelid.

"What were you thinking, Indie? You could have gotten yourself, not to mention Phoebe, killed."

He was right. Per usual, I'm too stubborn to admit it. "But I didn't."

Sean stares down at me, and two red welts dot his cheekbone where his eye patch normally rests.

"You saved me." Afraid my words might be misconstrued as grudging. "I haven't had the opportunity to say thank you." My fingertips touch the side of his left cheekbone gently. "By the way, how are those bee stings?"

His prickly exterior softens. "They're fine. The calendula cream your sister whipped up took the sting out of them."

"I'm glad you didn't get hurt." Self-consciously, my arms drop down to my side. "Is that patchouli I smell?"

"Yeah, at least I think so," Sean rubs the side of his neck. "Do you like it? Serena says it's supposed to relieve stress."

It makes you smell delicious, I almost say aloud. "I do. It's one of my favorite scents." Then, addressing the elephant in the alley, I say, "Perhaps we could agree the case was—a collaborative effort?"

"Perhaps," Sean says. He smiles. "But I don't want you getting the idea you're some kind of trained detective or anything."

"Oh, believe me, I don't think that." *Uh-uh,* not at all. Just in case, I cross my fingers behind my back on both hands.

"Good, I'm glad to hear it. So, you want to discuss the case over tea and cookies?"

"Vegan ones?"

"If you insist," Sean sighs.

"God, no, those things are awful. How about we compromise?"

"Yeah, they are pretty awful," Sean laughs aloud. "Does that mean I get to choose the cookies?"

I punch him playfully in the bicep, taking note of the rock-hard muscles. "No, Interim Sheriff Riordan, it means I choose the cookies and you get to pay the bill."

Sean laughs. "That's not exactly what I call compromise. But, okay—Indigo Bleu, you better watch out—I might get used to bargaining with you."

"That was my plan all along."

About the Author

R ebecca O'Bea is a writer, blogger, and author. *Hive and Seek* is the first installment of the **Backyard Beekeeping Mystery** series. When Rebecca is not busy writing, she lives with her husband on her small horse farm in Kansas, tending to a cadre of animals, including horses, chickens, two very spoiled dogs, and a barn cat who is convinced he is the leader of the pack, and of course, her hardworking honeybees.

If you enjoyed this book, please give us a review, and look for my upcoming release of Swarm and Sabotage, the second installment in The Backyard Beekeeping Mysteries, soon to Kindle.

Glossary of Beekeeping Terms

Glossary of Beekeeping Terms

Absconding swarm: the complete abandonment of a beehive by the adult population. Causes such as disease, lack of food, and other unfavorable conditions may vary.

Apiary: a place where bee colonies are kept and managed, also known as the bee yard.

Apiarist: another name for a beekeeper.

Apiculture: the science of raising honeybees

Apis: the genus of honeybees

Apis mellifera: the scientific name for the European honeybee most commonly cultivated in North America.

Bee gloves: cuffed and ventilated medium to long gloves generally made of leather and canvas designed to protect the beekeeper from stings, sticky honey, and propolis.

Bee Space: The scientific space of 3/8 inch allows for the buildout of comb and propolis. The spacing in the Langstroth Hive is based on this principle.

Bee Suit: a pair of coveralls, usually white, and heavy cotton weight, often with an attached veil. Designed to protect the beekeeper from stings.

Beeswax: the wax produced in the abdomen of adult female honey bees used to build out comb. Bees consume between 3-7 pounds of honey in order to produce 1 pound of wax.

Brood: the development of young bees within the three stages of egg, pupae, and larvae have yet to emerge from their cells.

Brood Chamber: the deep wooden box where the brood is raised. The *Langstroth* hive typically contains two deep brood chambers containing nine to ten moveable frames.

Capped honey: honey that is sealed by a layer of wax.

Carniolan Honey Bee: a strain of honeybees from Europe known for their gentle nature and ease of working with.

Colony: a group of bees that live as a social unit that contains a queen, thousands of worker bees, and drones.

Drone: a male bee. The drone population expands through the spring and summer before being eliminated from the hive in the late fall.

Extractor: a cylindrical barrel used to remove honey from individual frames by the use of centrifugal force without destroying the comb.

Glyphosates: common systemic herbicide absorbed through the plant leaves and minimally through the roots used to kill weeds and other commonly invasive weeds. Known to harm the microbiome in the guts of honeybees and other pollinators.

Honey: a viscous mixture of sucrose and fructose that bees produce from the nectar of flowers.

Insecticide: a chemical application that kills insects on contact.

Langstroth Hive: Patented by L.L. Langstroth, a Philadelphia reverend in 1852, first using the concept of "bee space" and the moveable frame hive. The hive generally consists of two "deep" brood chambers atop a bottom board with a hive entrance, an inner cover, and an outer cover.

Moisture content: the percentage of water should be no greater than 18.6 percent to minimize honey's crystallization.

Nectar: the sweet liquid found primarily in the flowers of a plant.

Nucleus or 'Nuc" hives: between two and five frames of comb and brood used to form a new hive. A fertile queen is added.

Package Bees: the acquisition of two and five pounds of adult bees with a fertile queen purchased in a wood and wire box to establish a new colony. A fertile queen is separated along with a few attendants in a small "queen cage." The queen acclimates to the worker bees by distributing pheromones throughout the hive. This process usually occurs over one week before she is released to begin laying eggs.

Pollen: dust-sized granules found in the anther of the male germ cell of flowering plants. The essential form of protein used to raise brood in a bee colony.

Propolis: a sticky substance formed by bees masticating sap or resin used to fortify and waterproof bee entrances and seal cracks in bee hives.

Queen: a female bee larger than an adult worker bee and able to lay fertilized eggs. Essential for the survival of the beehive.

Queen excluder: a slim wire board placed between the brood chambers and the honey supers that prevents the passage of the queen bee from laying eggs but allows the worker bees to store honey. Usually much narrower than the deep chambers below but capable of holding 30-50 pounds of honey.

Royal jelly: the thick white liquid secreted from the gland of a worker bee used to feed developing larvae and the queen bee.

Smoker: a device used to generate smoke in order to calm bees by triggering the autonomic response of "fight or flight." The handheld stainless steel can uses attached bellows to stoke the smoke upward via a hinged top and nozzle. The can is in a wireframe with a hook so the beekeeper can hold it without being burned by the heat.

Social insects: Insects that live within a family structure with the characteristics of cooperative work distribution, brood care, and reproductive ability to ensure the longevity of future generations.

Swarm: when the hive divides by the percentage of adult females, drones and the queen leave the hive to establish a new colony. Swarming season usually occurs in Spring when the hive rapidly builds out its population.

Varroa mite: an external parasite common to honeybees that causes anomalies in the growth and anatomy of the developing honeybee. Generally thought to be one of the causes of colony collapse.

Wax moth: a common pest whose larvae drill through the wax cappings on bee frames in search of food, destroying the bee larvae. They leave behind a tell-tale sign of spider-like webbing and can rapidly destroy a hive if not controlled.

Worker bee: an adult female bee unable to reproduce. Worker bees perform all the labor in the hive, with the exception of egg-laying. Their lifespan typically lasts 5-6 weeks during the summer months, with the adult worker moving from task to task within the hive as they age.

www.ingramcontent.com/pod-product-compliance
Lightning Source LLC
Chambersburg PA
CBHW051958220626
47052CB00004B/1001